BORANES AND METALLOBORANES
Structure, Bonding and Reactivity

ELLIS HORWOOD SERIES IN INORGANIC CHEMISTRY

Series Editor: J. BURGESS, Department of Chemistry, University of Leicester

Inorganic chemistry is a flourishing discipline in its own right and also plays a key role in many areas of organometallic, physical, biological, and industrial chemistry. This series is developed to reflect these various aspects of the subject from all levels of undergraduate teaching into the upper bracket of research.

BORANES AND METALLOBORANES
Structure, Bonding and Reactivity

CATHERINE E. HOUSECROFT B.Sc., Ph.D.
Fellow of Newnham College *and* Royal Society Research Fellow
University Chemical Laboratories, Cambridge

ELLIS HORWOOD LIMITED
Publishers · Chichester

Halsted Press: a division of
JOHN WILEY & SONS
New York · Chichester · Brisbane · Toronto

.3825115

CHEMISTRY

First published in 1990 by
ELLIS HORWOOD LIMITED
Market Cross House, Cooper Street,
Chichester, West Sussex, PO19 1EB, England
The publisher's colophon is reproduced from James Gillison's drawing of the ancient Market Cross, Chichester.

Distributors:

Australia and New Zealand:
JACARANDA WILEY LIMITED
GPO Box 859, Brisbane, Queensland 4001, Australia

Canada:
JOHN WILEY & SONS CANADA LIMITED
22 Worcester Road, Rexdale, Ontario, Canada

Europe and Africa:
JOHN WILEY & SONS LIMITED
Baffins Lane, Chichester, West Sussex, England

North and South America and the rest of the world:
Halsted Press: a division of
JOHN WILEY & SONS
605 Third Avenue, New York, NY 10158, USA

South-East Asia
JOHN WILEY & SONS (SEA) PTE LIMITED
37 Jalan Pemimpin # 05–04
Block B, Union Industrial Building, Singapore 2057

Indian Subcontinent
WILEY EASTERN LIMITED
4835/24 Ansari Road
Daryaganj, New Delhi 110002, India

© 1990 C.E. Housecroft/Ellis Horwood Limited

British Library Cataloguing in Publication Data
Housecroft, Catherine E.
Boranes and metalloboranes.
1. Boron hydrides
I. Title
546'.6712

Library of Congress Card No. 89–29980

ISBN 0–7458–0459–4 (Ellis Horwood Limited)
ISBN 0–470–21671–9 (Halsted Press)

Typeset in Times by Ellis Horwood Limited
Printed in Great Britain by The Camelot Press, Southampton

To Edwin, with love

Table of Contents

Preface

The aim of this book is to introduce the reader to the chemistry of compounds containing boron and hydrogen (boranes), and to investigate the ways in which these compounds interact with metal atoms. Although the element boron is a neighbour of carbon in the Periodic Table, the difference of just one electron between the two elements results in major changes in their chemistry. Carbon forms a myriad of compounds in which the bonding is readily described in localized terms. Even in aromatic or conjugated systems, allocation of electrons is not problem. On the other hand, a boron atom starts life with only three valence electrons. This seems a poor state of affairs for a non-metallic element. However, the chemical literature is proof enough that, somehow, the boron atom makes the most of it. *Boranes and Metalloboranes: Structure, Bonding and Reactivity* endeavours to describe how this is achieved in some of the inorganic compounds formed by boron.

The text is aimed primarily at an undergraduate audience, but I hope that it may also serve as a useful reference book for all those with an interest in the chemistry of boranes and metalloboranes. Although I have attempted to describe most topics in this book from first principles, I have left the theories of the spectroscopic techniques discussed in Chapter 3 to other authors.

I should like to acknowledge those people who, in one way or another, directly or indirectly, have contributed towards the writing of this book. I am deeply grateful to Professors Ken Wade and Tom Fehlner whose guidance through my graduate and postdoctoral years set the scene for my further research interests in the field of transition metal–main group element chemistry. The members, past, present and future, of my research group obviously play a vital rôle in firing my enthusiasm as they 'boldly go where (we hope!) no chemist has gone before'. The undergraduates whom I have the pleasure of teaching have certainly provided me with the motivation to write this book; I hope that a few of their questions are answered herein. My thanks are due to all my colleagues in Cambridge for their friendship and for their many and varied discussions and help with proof-reading, and especially to my husband, Edwin Constable, who is a constant source of inspiration and encouragement.

Cambridge, 1989   Catherine E. Housecroft

1

Introduction

1.1 THE POSITION OF BORON IN THE PERIODIC TABLE

Boron is the first element in Group IIIB of the Periodic Table (Fig. 1.1.1). It is located just above that so-called 'diagonal line' which separates metals from non-metals. We might well expect the chemistry of boron to be unique amongst its Group IIIB congeners as this element is the only member of the group which is 'on the side of' the non-metals. This is, indeed, the case as, I hope, the reader will soon appreciate.

| I | II | | | | | | | | | | | III | IV | V | VI | VII | VIII |
A	A											B	B	B	B	B	
H																	He
Li	Be											B	C	N	O	F	Ne
Na	Mg											Al	Si	P	S	Cl	Ar
K	Ca	Sc	Ti	V	Cr	Mn	Fe	Co	Ni	Cu	Zn	Ga	Ge	As	Se	Br	Kr
Rb	Sr	Y	Zr	Nb	Mo	Tc	Ru	Rh	Pd	Ag	Cd	In	Sn	Sb	Te	I	Xe
Cs	Ba	La	Hf	Ta	W	Re	Os	Ir	Pt	Au	Hg	Tl	Pb	Bi	Po	At	Rn
Fr	Ra	Ac															

Ce	Pr	Nd	Pm	Sm	Eu	Gd	Tb	Dy	Ho	Er	Tm	Yb	Lu
Th	Pa	U	Np	Pu	Am	Cm	Bk	Cf	Es	Fm	Md	No	Lr

Fig. 1.1.1 — The Periodic Table.

1.2 DEFINITIONS

We shall begin our journey through the chemistry of the boron hydrides and their metal-containing derivatives by defining a few terms which will be in constant use throughout this book. The classes of compound named in the title of this book are *boranes* and *metalloboranes*. A *borane*, strictly, contains only the elements boron and hydrogen, and is thus a boron hydride, just as a hydrocarbon is a carbon hydride. A *metalloborane* is a derivative of boron hydride in which the borane unit is attached to a metal (main group, transition, lanthanide, or actinide) containing fragment. Metalloborane compounds are a boron chemist's answer to organometallic complexes.

The electronic configuration of the boron atom is $1s^2 2s^2 2p^1$. In any discussion of the inorganic chemistry of boron, the term *electron deficient* usually appears. This may refer to the fact that when boron forms three covalent 2-centre 2-electron bonds (to satisfy its valency), the valence shell only contains six electrons. More generally, the term is applied to compounds of boron which apparently possess too few bonding electrons to hold themselves together in a coventional manner, i.e. it is not possible to describe the bonding in terms of localized 2-centre 2-electron covalent bonds. For instance, the simplest case would be diborane(6), B_2H_6, which has twelve valence electrons available and yet appears to form eight bonds (Fig. 1.2.1). The bonding in such a compound may be adequately described by the use of delocalized 3-centre 2-electron B–H–B bridges as will be illustrated in Chapters 2 and 6. In this way the boron atoms make the most of a poor situation, and the bonding electrons do rather more work than might otherwise be required of them. Of course, a consequence of this is that a bridging B–H interaction, (i.e. half of the complete B–H–B bridge), is weaker than a terminal B–H interaction.

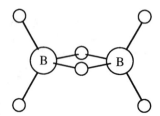

Fig. 1.2.1 — The structure of diborane(6), B_2H_6.

This leads us directly to a problem: if a compound is electron deficient, and the concept of localized 2-centre 2-electron bonds is inadequate, how should we represent the structure of a boron-containing compound? For example, in Fig. 1.2.1, should each line drawn between two adjacent atoms in the B_2H_6 molecule be termed a *bond*? The answer to this last question is no. A line drawn in a structural representation of a molecule in this book certainly indicates an interatomic *inter-action*, but it does not necessarily imply what the reader might consider to be a *bond*. This distinction will be explained in detail in Chapters 2 and 6.

Synonymous with many borane compounds is the word *cluster*. This description comes about because, in order to achieve efficient interatomic bonding, boron atoms often aggregate together, i.e. form clusters of atoms. Fig. 1.2.2 illustrates the structure of the anion $[B_{12}H_{12}]^{2-}$ and, for comparison, a unit of the lattice of the α-rhombohedral allotrope of elemental boron. The atoms of the anion define an icosahedral cluster. This particular cluster geometry is quite common in boron chemistry, and may be recognized as a structural unit in α-rhombohedral boron. This allotrope possesses an extended array of hexagonally close packed icosahedra, Fig. 1.2.2(b).

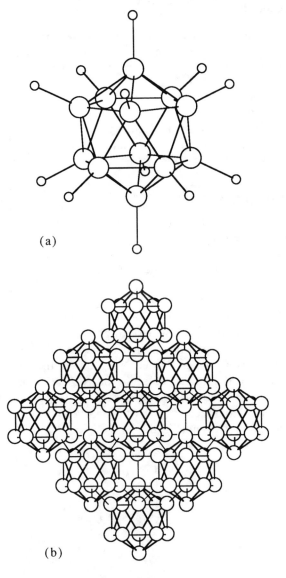

(a)

(b)

Fig. 1.2.2 — (a) The structure of the anion $[B_{12}H_{12}]^{2-}$, and (b) a unit of the lattice of α-rhombohedral boron.

Thinking in terms of 3-dimensional cluster frameworks again emphasizes the problems which we shall encounter when attempting to rationalize the way in which each boron atom uses its valence electrons. As the cluster becomes larger and more complex, the boron atom appears to indulge in more bonding interactions. The concept of electron deficiency reaches untold proportions! As we shall discover, simple hybridization schemes are very soon pushed to their limits when we try to describe the bonding in cluster molecules. Hence, it is advantageous to use a full molecular orbital treatment, and this concept will be introduced in Chapter 2 and discussed in more detail in Chapter 6. The list of abbreviations in Table 1.2.1 will be useful for the reader to note.

Table 1.2.1 — Abbreviations used in the text, and referring to a molecular orbital treatment of bonding

AO	Atomic orbital
MO	Molecular orbital
LGO	Ligand group orbital
HOMO	Highest occupied molecular orbital
LUMO	Lowest unoccupied molecular orbital

1.3 WHY STUDY BORANES?

A boron atom possesses only three valence electrons, and so what might we expect to be the predominant chemistry exhibited by this element? The first three ionization energies of boron are 801, 2427 and $3660\,kJ\,mol^{-1}$ respectively. If boron were to form an ionic compound, the lattice energy of the product would have to more than offset the net energy required to form the $B^{3+}(g)$ ion. As this is never the case, the chemistry of boron is one of covalent bond formation. So, with three available electrons we might anticipate that the boron atom will form three covalent bonds, and then, perhaps, look for a Lewis base to provide an extra pair of electrons to complete the $2s^22p^6$ valence configuration of boron. This expectation is indeed realized in the many compounds formed by boron. In the adduct $R_3N.BH_3$, R = alkyl, aryl; Fig. 1.3.1), the amine molecule forms a dative bond to the otherwise unstable BH_3 molecule. In the cyclic borazine, $B_3N_3H_6$ (an inorganic analogue of benzene), each boron atom receives an extra pair of electrons via a coordinate bond from an adjacent nitrogen atom (Fig. 1.3.2).

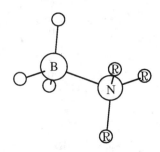

Fig. 1.3.1 — Structure of $R_3N.BH_3$.

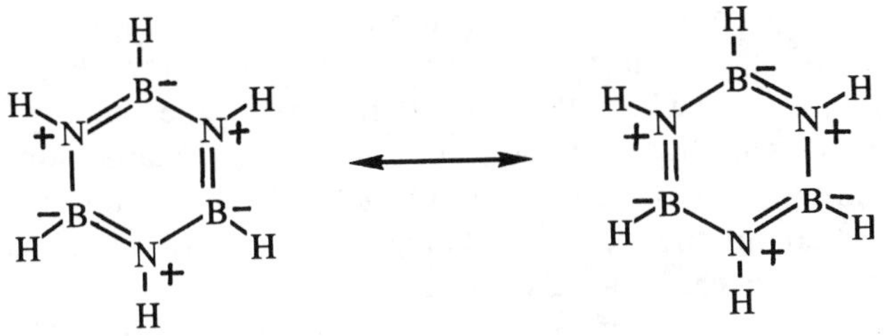

Why study boranes?

Fig. 1.3.2 — Structure of borazine.

Although the inorganic chemistry of boron is certainly dominated by its thirst for more electrons, the real beauty of this element lies in the variety and novelty of the ways in which it achieves its goal. Adduct formation is only the beginning. Two boron hydrides are shown in Fig. 1.3.3. The diagrams give a taste of some of the fascinating and complex structures adopted by those compounds. This originality of structure is as good a reason as any to study the boranes. Over the past decade, the number of homonuclear and heteronuclear clusters reported in the chemical literature has increased dramatically. Initially, the boranes led the way, showing a regular pattern in their structural behaviour. Each molecular shape may be rationalized in terms of a

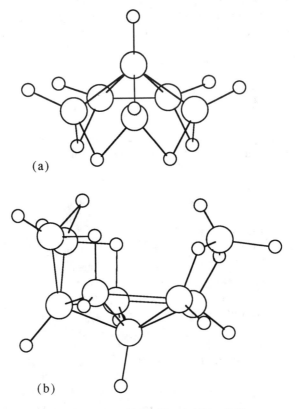

(a)

(b)

Fig. 1.3.3 — Structures of (a) B_6H_{10}, and (b) n-B_9H_{15}.

regular polyhedron or part thereof. The introduction of transition metal fragments adds a new dimension, not only in structural variation, but in bonding descriptions as well. The problems faced when attempting to rationalize the bonding in the boranes and their derivatives are further reasons for making them a target for investigation.

The structure of a molecule is often confirmed by means of a diffraction technique. Most commonly, X-ray crystallography is used for the analysis of solid-state materials. However, growing suitable crystals is not necessarily easy, as every researcher knows! Therefore, if an element or compound lends itself to one or more spectroscopic techniques, the benefits are enormous. Boron possesses not just one, but two, isotopes, ^{10}B and ^{11}B, with non-zero nuclear spin, and thus these nuclei are suitable for nuclear magnetic resonance spectroscopy. The introduction of an 'exotic nucleus' (i.e. something other than the more commonly observed ^1H, ^{13}C, ^{19}F and ^{31}P nuclei) to the NMR spectrometer is a further reason for making the study of boron-containing compounds an exciting prospect.

Finally, if the structures of the boranes are novel, then we should expect their chemistry to be unusual, and, maybe, unpredictable! Does one require any more reasons for studying a material?

1.4 NOMENCLATURE

Listed below are some common terms used in naming boron hydride compounds. The name for a boron hydride indicates the number of boron and hydrogen atoms: a Latin prefix gives the number of B atoms and an Arabic number in parentheses shows the number of H atoms. For anions, the charge is indicated in parentheses.

borane	boron hydride
hydroborate	boron hydride anion
mono-	one
di-	two
tri-	three
tetra-	four
penta-	five
hexa-	six
hepta-	seven
octa-	eight
nona-	nine
deca-	ten
unideca-	eleven
dodeca-	twelve
trideca-	thirteen
tetradeca-	fourteen
icosa-	twenty

Examples:		
	Pentaborane(9)	B_5H_9
	Pentaborane(11)	B_5H_{11}
	Octaborane(12)	B_8H_{12}
	Octahydrooctaborate(2−)	$[B_8H_8]^{2-}$

The American Chemical Society has outlined its accepted nomenclature for boranes, and this is commonly used throughout the literature: *Inorg. Chem.*, 1968, **7**, 1945. Numbering schemes for some of the borane cages are given in Figs 7.1.5, 7.2.1, 7.3.1 and 7.3.2.

2

The Lewis acidity of BH₃

2.1 THE STRUCTURE AND BONDING REQUIREMENTS OF BH₃

The ground state electronic configuration of the boron atom is $1s^2 2s^2 2p^1$ and is represented in Fig 2.1.1.

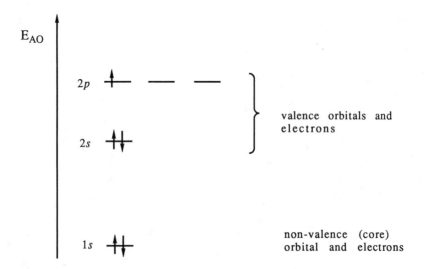

Fig. 2.1.1 — The ground state electronic configuration of B.

According to the valence shell electron pair repulsion (VSEPR) theory, use of the three valence electrons to form three covalent bonds should produce a trigonal planar molecule. Synonymous with trigonal planarity is an sp^2 hybridization scheme. One of the ground state $2s$ electrons is promoted to a $2p$ orbital, and subsequent mixing of the $2s$ with two $2p$ orbitals provides three hybrid orbitals of equal energy. This is illustrated on the left-hand side of Fig. 2.1.2.

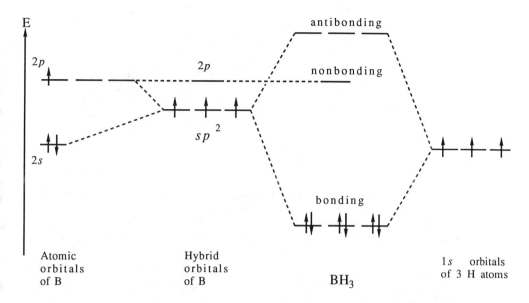

Fig. 2.1.2 — Bonding scheme for trigonal planar BH$_3$ using sp^2 hybridization.

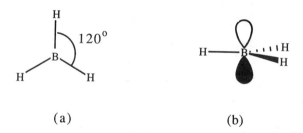

(a)   (b)

Fig. 2.1.3 — Structure of monomeric BH$_3$, viewed (a) from above, and (b) from the side showing the orientation of the empty $2p$ atomic orbital.

Monoborane, BH$_3$ (Fig. 2.1.3), is the simplest borane known. Although its existence in the gas phase has been established, BH$_3$ is not an easy molecule to study experimentally. The reason for this is seen in Fig. 2.1.3(b); the boron atom in BH$_3$ carries an empty $2p$ orbital and is therefore extremely susceptible to attack by nucleophiles. This is demonstrated in the molecular orbital diagram shown in Fig. 2.1.2. Each of the three sp^2 hybrid orbitals interacts with a hydrogen atom $1s$ orbital to generate three B–H bonding MOs (σ_{BH}) (Figs 2.1.2 and 2.1.4), and three B–H antibonding MOs (σ^*_{BH}). The formation of the three B–H bonds does not supply a complete octet of electrons around the boron atom. The $2p$ orbital remains empty and is nonbonding with respect to the BH$_3$ molecule. Thus, monoborane is willing to accept a pair of electrons in order to attain an inert gas configuration, and the nonbonding $2p$ orbital is therefore a site of nucleophilic attack.

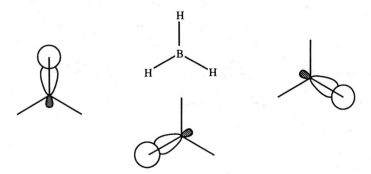

Fig. 2.1.4 — Representation of the three bonding orbitals in BH$_3$ using an *sp^2* hybridization scheme.

At this point, we must emphasize that hybridization schemes are used *for our convenience*. For instance, an *sp^2* hybridization scheme is a convenient way of explaining the formation of the three equivalent bonds in a BH$_3$ molecule. When a covalent molecule forms from its constituent atoms, bonding is achieved by the use of *atomic orbitals*. Combinations of these orbitals generate molecular orbitals, and the complete bonding description of the final molecule is obtained from an *overall* consideration of the filled molecular orbitals. Consider monoborane again. Fig. 2.1.5

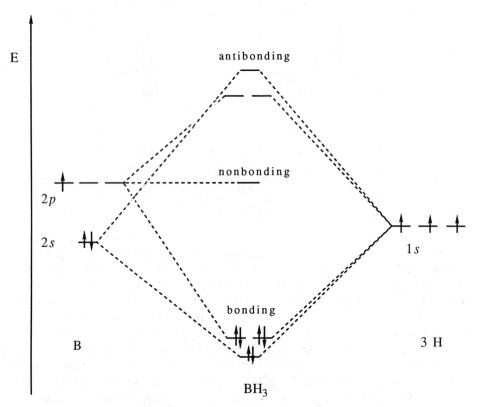

Fig. 2.1.5 — Molecular orbital diagram for the formation of BH$_3$ from an atomic orbital basis set.

shows the interactions between the atomic orbitals of the boron atom and the three $1s$ orbitals of the three hydrogen atoms to generate a trigonal planar molecule of BH_3, and, thus, a bonding scheme for the molecule is constructed by using an *atomic orbital basis set*. We observe that there are three filled bonding orbitals, one unfilled nonbonding orbital, and three empty antibonding orbitals. To this extent, Fig. 2.1.5 resembles the right-hand side of Fig. 2.1.2. However, the bonding MOs in Fig. 2.1.5 are not of equivalent energy, and since BH_3 possesses three equivalent B–H bonds, the non-equivalence of the bonding orbitals may appear to be a problem. Let us re-examine the nature of the orbitals in an attempt to rationalize this apparent discrepancy.

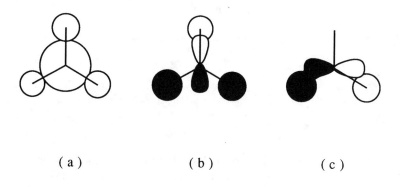

(a)   (b)   (c)

Fig. 2.1.6 — Schematic representations of the bonding molecular orbitals in BH_3.

The lowest lying MO in Fig. 2.1.5 is generated by the interaction of the boron $2s$ AO with the H atom $1s$ orbitals. An in-phase combination of these atomic orbitals is shown in Fig. 2.1.6(a). This single molecular orbital has bonding character along *each* B–H vector. The higher lying bonding MOs in Fig. 2.1.5 are degenerate, and arise from the interaction of two of the boron $2p$ AOs with $1s$ orbitals on the H atoms. Each $2p$ orbital possesses a nodal plane, and hence, in order to generate combinations of boron and hydrogen atomic orbitals which are B–H bonding, we must take out-of-phase combinations of hydrogen $1s$ orbitals. These are illustrated in Figs 2.1.6(b) and 2.1.6(c). In the latter diagram, one H atom is coincident with the nodal plane passing through the $2p$ orbital of the boron atom. As the $1s$ orbital has no nodal plane of its own, there is a mismatch in symmetries between this $1s$ orbital and the $2p$ orbital. The interaction is deemed to be *disallowed by symmetry*. Another way of viewing this phenomenon is to consider the consequences of allowing the interaction. This is illustrated in Fig. 2.1.7.

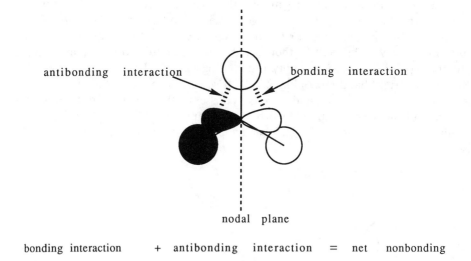

antibonding   interaction   bonding   interaction

nodal   plane

bonding interaction   +   antibonding   interaction   =   net   nonbonding

Fig. 2.1.7 — Schematic representation of a symmetry disallowed interaction between a
hydrogen 1s orbital residing on the nodal plane of a boron 2p orbital.

Although the 1s orbital overlaps favourably with one lobe of the 2p orbital, the other
half of the interaction is antibonding, and there is, therefore, no net orbital overlap.

Let us now return to the BH₃ molecule as a whole. The molecular orbitals in Figs
2.1.6 are clearly different from one another, unlike those illustrated in Fig. 2.1.4.
Nevertheless, we can still rationalize the existence of three equivalent B–H bonds as
follows. The lowest lying bonding MO presents no problem; it provides equal B–H
bonding character for each B–H interaction. Taken together, the degenerate set of
bonding MOs provides one *direct* B–H bonding interaction (Fig. 2.1.6(b)) and four
indirect interactions involving two of the B–H bonds (Figs 2.1.6(b) and (c)). Orbitals
(except s) possess vectorial properties, and, therefore, the combination of the four
indirect B–H orbital interactions will lead to two direct interactions. Hence, the
degenerate pair of filled MOs in BH₃ contribute a total of three direct, and
equivalent, B–H bonding interactions, and overall, the three filled MOs provide the
BH₃ molecule with three equivalent bonds. The conclusion, then, is that Figs 2.1.2
and 2.1.5 are really telling the same story.

Are we justified in allowing the atomic orbitals on the boron atom to hybridize
before bonding? Photoelectron spectroscopy is an experimental technique by which
the energies of molecular orbitals can be assessed. For methane, in which there are
four geometrically and chemically equivalent C–H bonds, the photoelectron spectrum exhibits *two* ionizations. One band is attributed to electron loss from the
degenerate MOs derived from overlap of the carbon atom $2p_x$, $2p_y$ and $2p_z$ AOs with
hydrogen atom 1s AOs. The second ionization is assigned to electron loss from the
MO derived from overlap of the carbon 2s with hydrogen 1s AOs. Hence, this
spectroscopic method implies that the four MOs in methane are *not* of equal energy.

Similarly, the bonding picture in BH$_3$ is more accurately represented by the MO diagram in Fig. 2.1.5 rather than Fig. 2.1.2. Nevertheless, it is certainly *convenient* for us to think in terms of hybridization, and so we shall, from now on, do this whenever it will make life easier for us.

In both the bonding schemes described above, one important feature of the BH$_3$ molecule emerged. The nonbonding 2p orbital on the boron atom is vacant, and can accept electrons. These electrons are provided by a Lewis base. Before considering such reactions, we shall discuss why it is that not all tricoordinate boron compounds are unstable with respect to nucleophilic attack.

Consider the family of boron trihalides, BX$_3$ (X = F, Cl, Br, I). Each of these compounds is monomeric and has a trigonal planar geometry. Each behaves as a Lewis acid, but the acid strength is significantly less than that of BH$_3$. Moreover, the Lewis acidity of the boron trihalides follows the sequence BF$_3$ < BCl$_3$ < BBr$_3$ < BI$_3$. This is, perhaps, initially surprising. The electronegativity of the substituents follows the trend F > Cl > Br > I, thereby implying that of the boron trihalides, BF$_3$ should be the most susceptible to the approach of a nucleophile. Why, then, is it in fact the *weakest* Lewis acid? The answer lies in the ability of the halides, and in particular the fluorine atom, to donate lone pair π-electron density into the empty 2p orbital on the central boron atom. This situation, using BF$_3$ as the example, is contrasted with that in BH$_3$ in Fig. 2.1.8. In BH$_3$, the hydrogen 1s orbitals have nothing more to offer once the B–H σ-bonds have been formed. On the other hand, having formed a B–F σ-bond, a fluorine atom still possesses three lone pairs of electrons, one of which resides in a 2p AO oriented parallel to the empty 2p orbital on the boron atom. Delocalization of π-electron density can occur over the three B–F bonds, thereby strengthening them, and stabilizing the BF$_3$ molecule. The same is true for BCl$_3$, BBr$_3$, and BI$_3$, but to an ever decreasing degree. Thus, in going from BF$_3$ to BI$_3$, the Lewis acidity is enhanced.

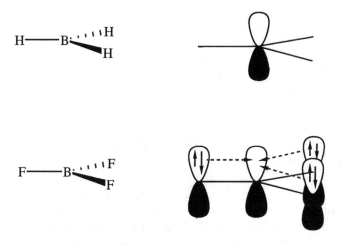

Fig. 2.1.8 — The availability of pπ–pπ bonding in BF$_3$ in contrast to BH$_3$.

2.2 ADDUCT FORMATION BY BH₃

We have shown in section 2.1 that the BH_3 molecule desires another pair of electrons, and a Lewis base is one obvious source. Solvents such as ethers, or ligands such as amines or phosphines readily coordinate to monoborane. Such is the stabilizing influence of these bases that, as a Lewis base adduct, BH_3 now becomes a viable chemical for laboratory use. For instance, thf.BH_3 (thf = tetrahydrofuran, C_4H_8O) is a commercial source of monoborane. The complex may be stored in thf solution in glass bottles, provided the atmosphere above the solution is inert (boron hydrogen bonds hydrolyse readily — see Chapter 7). Since monoborane is an important reducing and hydroborating agent, stabilization of the molecule is advantageous. Other Lewis base adducts of BH_3 which are available commercially include complexes with amines, phosphines, and thioethers. The relative stabilities of adducts, L.BH_3, vary according to the electron-donating ability of the Lewis base and with the steric properties of L. The ordering, according to L, for some common adducts is

$$PF_3 < CO < Me_2O < thf < C_4H_8S < Me_2S < py < Me_3N < Me_3P < H^-$$

Neither PF_3 nor CO is generally regarded as being a particularly efficient Lewis base; the electronegative fluorine atoms drain electron density away from the phosphorus atom, thereby rendering PF_3 a far poorer donor than, say, PMe_3. Even so, the strength of the Lewis acidity of monoborane permits the formation of adducts with even the weakest of Lewis bases. Competition between the ligands is observed in a series of displacement reactions. As the series above implies, the thf ligand is readily displaced by, for example, amines or trialkyl or triaryl phosphines. Some interesting reactions are possible, for example that of thf.BH_3 with a bidentate phosphine:

On forming an adduct, L.BH_3, the borane moiety loses its trigonal planar geometry. This is illustrated in Fig. 2.2.1. Once the boron $2p$ AO accepts a pair of electrons, it is no longer nonbonding with respect to the borane molecule. A change in hybridization must occur in order to accommodate the four bonding pairs of electrons. Thus, the borane undergoes a change from a trigonal planar (sp^2 B) to a tetrahedral (sp^3 B) geometry. The distortion is also readily understood in terms of the VSEPR theory. Repulsion between three pairs of bonding electrons in the case of free BH_3 leads to HBH bond angles of 120°, and repulsion between four pairs in the adduct, L.BH_3 produces bond angles at the boron atom close to 109.5°.

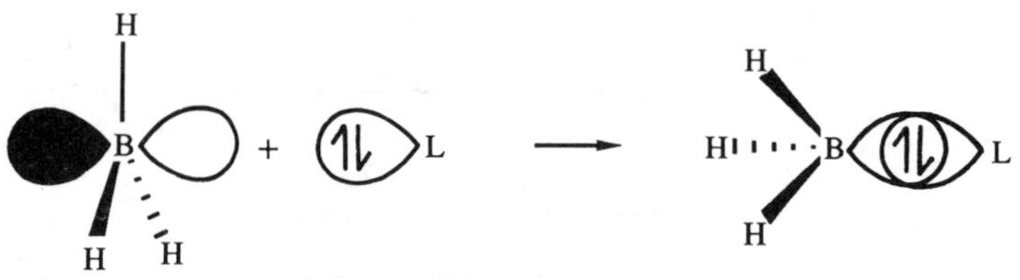

Fig. 2.2.1 — Change in geometry which accompanies formation of the L.BH₃ adduct.

2.3 DIMERIZATION BY BH₃

The ability of the BH₃ molecule to accept electrons from almost any source is emphasized by the fact that dimerization occurs in the absence of the external ligand. During dimerization, BH₃ functions both as a Lewis acid *and* as a Lewis base. The process may, at first glance, appear to be similar to the formation of Al₂Cl₆, since both B and Al are members of Group IIIB. However, a terminal chlorine atom in an AlCl₃ monomer is able to donate a lone pair of electrons into the empty $3p$ AO of an aluminium atom on an adjacent monomer, thereby creating a dative bond and an Al–Cl–Al bridge (Fig. 2.3.1). Each 2-centre bond is a conventional one in the sense that it possesses 2 electrons localized between two nuclei. This cannot be the case in the formation of diborane, since there are no lone pairs of electrons available. Instead, the B–H *bonding* electrons are made available for donation from one monomeric unit to the other. A pair of electrons which, in the monomer, binds two nuclei together is, in the dimer, delocalized over three nuclei (Fig. 2.3.2). Each H–B component of a 3-centre 2-electron bridge is weaker (and longer) than a terminal

Fig. 2.3.1 — Dimerization of AlCl₃ .

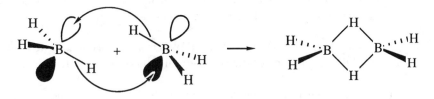

Fig. 2.3.2 — Dimerization of BH₃ .

B–H bond. Nonetheless, since each boron atom in the B₂H₆ molecule is surrounded by an octet of electrons, the situation is a favourable one and the best possible under the circumstances.

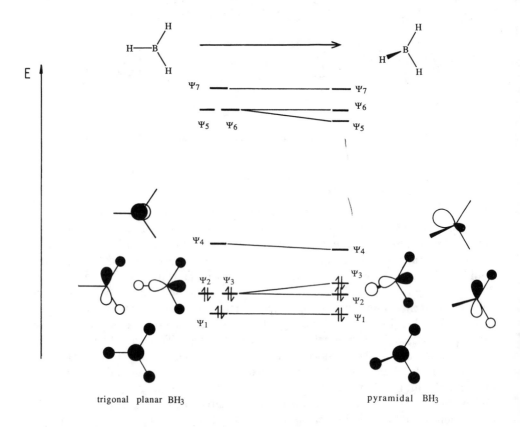

Fig. 2.3.3 — Perturbation of the MOs of trigonal planar BH$_3$ on distortion to pyramidal BH$_3$.

The dimerization of BH$_3$ is considered below in molecular orbital terms. Before the two molecules of BH$_3$ combine, each undergoes a notional distortion from a trigonal planar to a pyramidal geometry, thus providing a molecular geometry about each boron atom which mimics that observed experimentally in B$_2$H$_6$. Altering the molecular geometry of the monoborane fragment naturally affects its molecular orbitals as is shown in Fig. 2.3.3. Two significant changes occur:

(1) the degeneracy of the HOMOs in planar BH$_3$ is removed as one of the H atoms moves out of plane, and one of the B $2p$ orbitals adds out-of-plane character (ψ_3 in Fig. 2.3.3),

(2) the LUMO in planar BH$_3$ (originally a pure B $2p$ AO) adds $2s$ and in-plane $2p$ character and as a result is directed away from the pyramidal BH$_3$ molecule (ψ_4 in Fig. 2.3.3).

Thus, the HOMO and LUMO of the pyramidal BH$_3$ molecule are, respectively, a donor orbital with B–H bonding character, and an acceptor orbital centred on the boron atom.

E

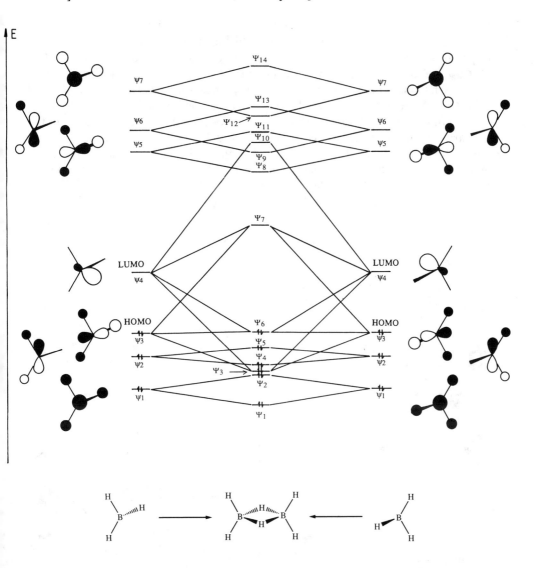

Fig. 2.3.4 — Molecular orbital correlation diagram for the dimerization of BH₃, beginning with pyramidal BH₃; note the mixing of ψ_3 and ψ_4 to generate B–H–B bridging orbitals.

Before we consider the orbital correlation diagram for the formation of B_2H_6 from two pyramidal BH_3 units illustrated in Fig. 2.3.4, let us review the formation of molecular orbitals for a diatomic molecule. In the simplest case, H_2, the H $1s$ AOs combine either in- or out-of-phase to give one bonding and one antibonding MO. In O_2, combinations are allowed between corresponding pairs of AOs on each oxygen atom to give σ-, π-, σ*-, and π*-MOs; mixing of the AOs is permitted if the symmetries of the AOs are compatible. Now consider the dimerization of BH_3. As the two pyramidal BH_3 fragments are brought together (Fig. 2.3.4), the lowest lying filled MOs (ψ_1 of each BH$_3$) can interact since their orbital characteristics are compatible. One bonding and one antibonding combination result, and these are, respectively, Ψ_1 and Ψ_2 in B$_2$H$_6$. Similarly, interaction between the second filled MOs, ψ_2, of the two BH$_3$ fragments generates interfragment bonding and antibonding MOs in B$_2$H$_6$. The donor and acceptor properties respectively of the HOMO and LUMO of each pyramidal BH$_3$ fragment are demonstrated by the fact that the LUMO of each one interacts with the HOMO of the other, and this results in the primary orbital interaction which stabilizes the B$_2$H$_6$ molecule. Significantly, these interactions, Ψ_3 and Ψ_6 (Fig. 2.3.5), generate two B–H–B bridges in the final B$_2$H$_6$ molecule. Hence, the HOMO of pyramidal BH$_3$ allows the molecule to function as a Lewis base, and the LUMO allows monoborane to function as a Lewis acid.

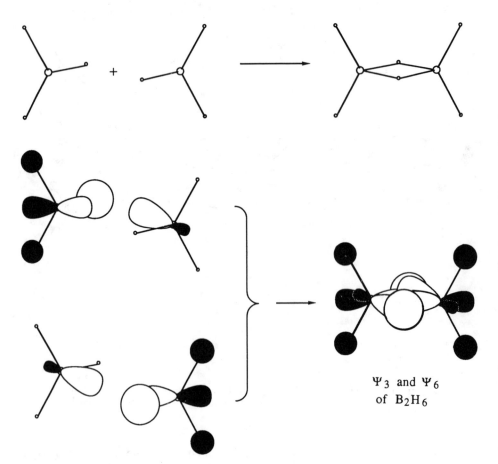

Ψ_3 and Ψ_6
of B$_2$H$_6$

Fig. 2.3.5 — Schematic representation of the Lewis acid–base interaction of 2 BH$_3$ molecules and consequent generation of Ψ_3 and Ψ_6 (both MOs have similar characteristics) in B$_2$H$_6$; two antibonding MOs, Ψ_7 and Ψ_{10}, also result.

Other bonding descriptions for B$_2$H$_6$ will be discussed in Chapter 6.

2.4 REACTIONS OF B₂H₆ WITH LEWIS BASES

A criterion for the formation of B_2H_6 from BH_3 is the absence of competing ligands. If a Lewis base, L, is present, it is energetically more favourable to form two equivalents of $L.BH_3$. This is not surprising, since the bonding capacity of the electrons in B_2H_6 is being stretched to their limit. If a molecule comprising all localized 2-centre 2-electron bonds (i.e. $L.BH_3$) can form in preference to the electron-deficient diborane, it will naturally do so.

The reaction of diborane with a Lewis base may follow one of several pathways, depending upon the nature of the nucleophile. Cleavage of the dimer into two moles of BH_3 is a common mechanism, and occurs with a wide range of bases. This pathway is known as *symmetrical cleavage* and is summarized in Fig. 2.4.1 for attack by trimethylamine. Analogous reactions take place with $(C_2H_5)_2O$, py, CO, PMe_3, PPh_3 and H^-. For reaction with the hydride ion, a non-competing solvent must be chosen. Under certain conditions, *asymmetrical* cleavage of the B_2H_6 molecule may be achieved; Fig. 2.4.2 illustrates the case for $L = NH_3$. The base must not be sterically demanding, or competition with the symmetrical cleavage pathway becomes significant. That asymmetrical cleavage has been achieved is evidenced by the formation of a 1:1 electrolyte. Choice of solvent is important; a basic medium favours symmetrical cleavage.

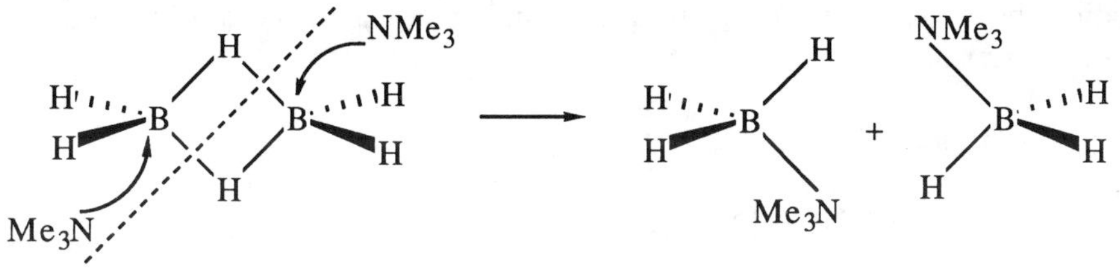

Fig. 2.4.1 — Symmetrical cleavage of B_2H_6 by trimethylamine.

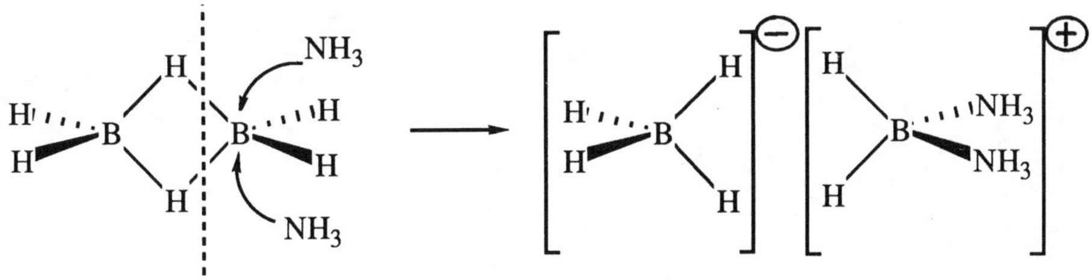

Fig. 2.4.2 — Asymmetrical cleavage of B_2H_6 by ammonia.

In both the symmetrical and the asymmetrical cleavage pathways, the first step in the reaction is probably the opening of one of the B–H–B bridges in the B_2H_6 molecule. Evidence for this comes from the formation of $[B_2H_7]^-$ in the reaction of B_2H_6 with H^- ion (Fig. 2.4.3). This ion may be viewed either as an adduct of B_2H_6 with H^-, or as an adduct of BH_3 with $[BH_4]^-$.

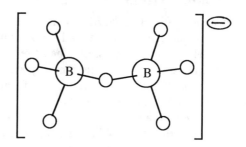

Fig. 2.4.3 — Structure of the $[B_2H_7]^-$ ion.

The above reactions of diborane with Lewis bases may be extended to the higher boron hydrides (see Chapter 7). Often, the base will remove a BH_3 unit from the reactant:

$$B_4H_{10}+2PMe_3 \rightarrow Me_3P.BH_3+Me_3P.B_3H_7$$

For these compounds, however, a fourth reaction pathway, that of simple deprotonation, becomes prevalent. Here, the incoming base abstracts a bridging hydrogen atom, for example:

$$B_5H_9+H^- \rightarrow [B_5H_8]^- +H_2$$

$$B_{10}H_{14}+NH_3 \rightarrow [NH_4][B_{10}H_{13}]$$

3

Physical methods applied to boranes and metalloboranes

3.1 TECHNIQUES AVAILABLE

The process of identifying a compound and elucidating its structure is similar to completing a jigsaw puzzle. A wide variety of spectroscopic and analytical techniques is available. One technique may provide a unique piece of information, whereas another may simply serve to confirm the conclusions drawn from the results of other analyses. Before we can confidently formulate and draw the molecular structure of a new compound, all the pieces of the puzzle must fit together, i.e. all the deductions made from studying the analytical and spectroscopic data must be consistent with one another.

The aim of this chapter is to illustrate to the reader the use of some of the more commonly available physical methods of analysis. The chapter does not set out to provide a comprehensive survey of physical methods, nor does it describe the theory and experimental details of any technique. Such information may be found elsewhere [1–4].

The applications of some physical techniques which are commonly employed by inorganic chemists are summarized in Fig. 3.1.1. Interplay of the methods is a key feature when one is analysing a newly synthesized compound. It might appear that an X-ray crystallographic study would independently and categorically establish both the identity and the structure of the compound. Thus, we might be tempted to assume that this one procedure is all that the chemist requires. Not so! A compound may not possess the same structure in solution or in the gas phase as it does in the solid state. This state of affairs is particularly true for boranes in solution, where hydrogen atoms may exchange rapidly between several sites. The borane is then said to exhibit *fluxional* behaviour. The same is true for metalloboranes, although here, additional fluxional processes involving the metal-attached ligands may also operate in solution. It should be noted that the observation of a fluxional process depends on the relationship between the rate of this process and the timescale of the experimental technique; e.g. for X-ray diffraction, the timescale is of the order of 10^{-18} s whereas for NMR spectroscopy it is slower, typically 10^{-9} to 10^{-1} s. It is also possible

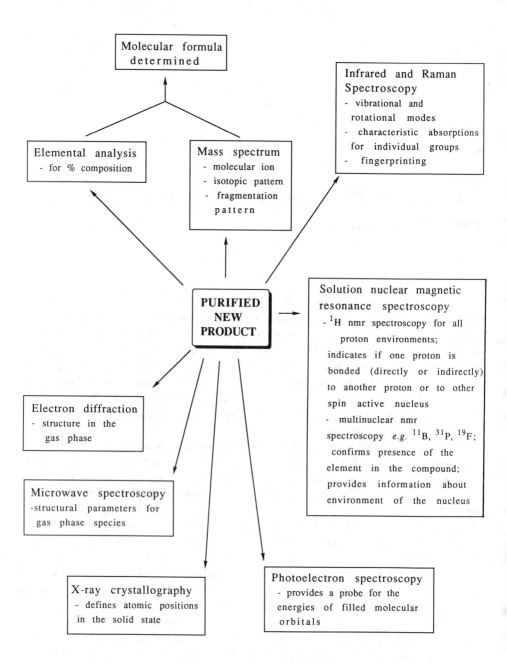

Fig. 3.1.1 — Commonly available physical techniques which may be used to formulate and elucidate the structure of an inorganic compound, including a borane or metalloborane.

that several isomers of a compound coexist in solution, but perhaps only one isomer is favoured in the solid state. This change is often attributed to the effects of packing forces between molecules in the crystal lattice. It should also be noted that X-rays are diffracted by electrons. Since a hydrogen atom possesses a single valence electron, there will be little residual electron density near to the nucleus of the hydrogen atom once covalent bond formation has occurred. X-ray diffraction data will always *underestimate* the internuclear separation for a bond involving a hydrogen atom. In the absence of direct hydrogen atom location, solution state ^{1}H nuclear magnetic resonance spectroscopy can provide information allowing the structural assignment to be completed.

In the following sections, we shall concentrate on the spectroscopic methods which are used most routinely by boron chemists, and then allude briefly to structural determination by diffraction methods.

3.2 MASS SPECTROMETRY

As a sample molecule, M, passes through a mass spectrometer, two principal processes occur. The first is ionization of M, and the second is fragmentation of the molecule. Ionization normally occurs according to the equation

$$M(g) + e^- (\text{high energy} \approx 70\,eV) \rightarrow [M^{\bullet+}](g) + 2e^- (\text{low energy})$$

The formation of multiply charged ions is far less common than the formation of singly charged ones, simply because the energy required for their formation is usually greater than that available. The ion $[M^{\bullet+}]$ is known as the molecular ion, and is derived directly by simple electron loss from a molecule of the compound being analysed. Some molecular ions will pass through the spectrometer and be detected intact, whereas other ions will fragment. For a given molecule, the fragmentation pathway is characteristic. Each molecular or fragment ion is characterized by a mass/charge ratio, m/e, and since e is usually unity, a value of m/e provides the mass of the ion directly. A mass spectrum is a plot of the ion current against m/e.

We shall now consider the consequences of a nucleus possessing more than one naturally occurring isotope. Elements such as F, P, Co, Rh and Au are monotopic. Thus, the presence of an ion such as Au^+ gives rise to just one line in the mass spectrum. Hydrogen may be regarded as possessing a single isotope since the ^{1}H nuclide occurs with an abundance of 99.985%. On the other hand, the naturally occurring minor isotope of carbon ^{13}C (1.11%), may *not* be ignored. Although for a single carbon atom, the mass spectral peak arising from the $^{13}C^+$ ion, (P+1), is negligible, for a molecular ion containing several carbon atoms, the (P+1) peak becomes significant. The calculated peak intensities for the ion C_n^+, where $n = 1, 2, 6, 12$ and 20, are given in Table 3.2.1, and the results are shown diagrammatically in Fig. 3.2.1. For each C_n^+ sample, the ions $(^{12}C_n)^+$, $(^{12}C_{n-1}{}^{13}C)^+$, $(^{12}C_{n-2}{}^{13}C_2)^+, \ldots$ $(^{12}C^{13}C_{n-1})^+$, $(^{13}C_n)^+$ will be present. The (P+1) peak is attributable to the presence of the $(^{12}C_{n-1}{}^{13}C)^+$ ion. The probability of forming ions containing several ^{13}C atoms is obviously far lower than the probability of observing the $(^{12}C_{n-1}{}^{13}C)^+$ ion. The characteristic observation of a (P+1) ion is noted later in the discussion of the mass spectrum of $Ru_3(CO)_{12}$ illustrated in Figs 3.2.7 and 3.2.8.

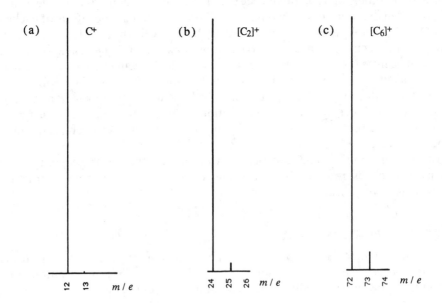

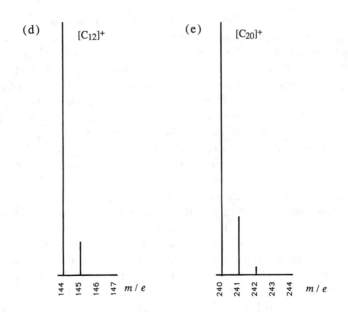

Fig. 3.2.1 — Isotopic patterns for the molecular ions of (a) C^+, (b) C_2^+, (c) C_6^+, (d) C_{12}^+, (e) C_{20}^+.

Table 3.2.1 — Calculated mass spectral peak intensities, given as a percentage, for the ion C_n^+

n	$(^{12}C_n)^+$	$(^{12}C_{n-1}{}^{13}C)^+$	$(^{12}C_{n-2}{}^{13}C_2)^+$	$(^{12}C_{n-3}{}^{13}C_3)^+$	$(^{12}C_{n-4}{}^{13}C_4)^+$
1	98.89	1.11	—	—	—
2	97.79	2.20	0.01	—	—
6	93.52	6.30	0.18[a]	—	—
12	87.46	11.78	0.73	0.03[a]	—
20	79.99	17.96	1.91	0.13	0.01[a]

[a]Peak intensities beyond this point are negligible.

Many elements possess several isotopes, each being present in significant abundance. Boron is a prime example; ^{10}B has an abundance of 19.7%, and ^{11}B of 80.3%. A mass spectrum of the B^+ ion is represented in Fig. 3.2.2(a), where the 1:4 relative intensities of the peaks at $m/e = 10$ and 11 are clearly visible. Similarly, chlorine exhibits two isotopes, ^{35}Cl (75.53%) and ^{37}Cl (24.47%) (Fig. 3.2.2(b)). Ruthenium is an example of a multi-isotopic element. It possesses seven naturally occurring nuclides as indicated by the mass spectrum shown in Fig. 3.2.2(c).

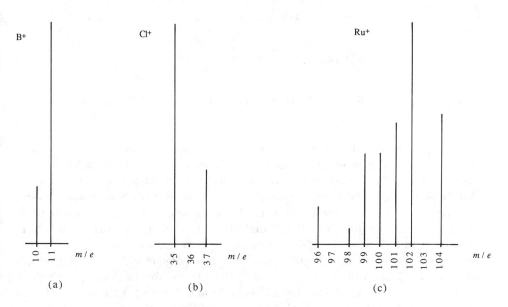

Fig. 3.2.2 — Representations of mass spectra of (a) B^+, (b) Cl^+, and (c) Ru^+ ions.

Isotope patterns are diagnostic. If an ion contains several atoms, the peaks in the mass spectrum will reflect the presence of different isotopes. Thus for a $[BH]^+$ ion, there are peaks due to $[^{10}B^1H]^+$ and $[^{11}B^1H]^+$ (Fig. 3.2.3(a)), and for a $[Cl_2]^+$ molecular ion, there are three peaks due to the ions $[^{35}Cl^{35}Cl]^+$, $[^{35}Cl^{37}Cl]^+$ or $[^{37}Cl^{35}Cl]^+$, and $[^{37}Cl^{37}Cl]^+$ (Fig. 3.2.3(b)). Note that the peak pattern for the $[BH]^+$ ion is dominated by the isotopic distribution due to the boron atom, since hydrogen is essentially monotopic.

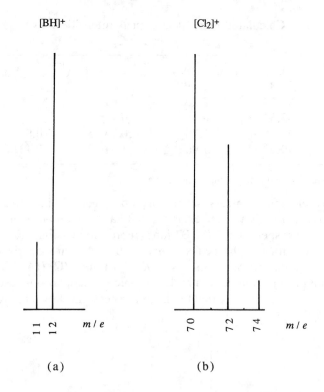

Fig. 3.2.3 — Representations of mass spectra of (a) $[BH]^+$ and (b) $[Cl_2]^+$ ions.

The mass spectra described above are recorded at low resolution, and the ratio *m/e* is determined as an integral value. Often, knowing an integral mass for an ion is insufficient to allow the unambiguous assignment of a molecular formula. Isotopic distribution patterns may aid identification, but in the absence of such information, *high resolution* mass spectrometry is used. This would, for example, allow us to distinguish between the ions $[^{11}B^1H]^+$ and $^{12}C^+$. The exact masses of the ^{11}B and ^1H nuclei are 11.009305 and 1.007825 a.m.u. respectively, whereas that of the ^{12}C nucleus is 12.000000 a.m.u. by definition. Hence, the high resolution mass spectrum of the $[^{11}B^1H]^+$ ion will exhibit a peak at 12.017130, and the ion is therefore readily distinguished from $^{12}C^+$. Similarly, a $[^{10}B^1H]^+$ ion is distinct from a single $^{11}B^+$ ion since the exact masses are 11.020763 and 11.009305 respectively.

Let us now examine some specific case histories, and define the information available from mass spectra, as well as the complications which may arise.

1.   Diborane(6), B_2H_6, and tetraborane(10), B_4H_{10}

A calculated group (an *envelope*) of peaks for the molecular ion in the mass spectrum of diborane, B_2H_6, is illustrated in Fig. 3.2.4(a). The isotopic distribution resembles that observed for a B_2^+ ion. (Fig. 3.2.4(b)). This similarity in pattern underlines the fact that, to a first approximation, the contribution made by the hydrogen nuclei to the isotopic pattern may be ignored. The experimentally determined [5] mass spectrum of B_2H_6 is illustrated in Fig. 3.2.4(c). Two features are immediately apparent. Firstly, fragmentation to $[BH_3]^+$ occurs. Secondly, the isotopic distribution for the parent ion does not coincide with the simulated spectrum shown in Fig. 3.2.4(a). The first feature is to be expected, since cleavage of the B_2H_6 molecule will occur at the 3-centre 2-electron B–H–B bridges. Note, however, that since the intensity of the peak at $m/e=14$ is low, the $[BH_3]^+$ ion itself is apparently not particularly stable. Loss of hydrogen atoms occurs to produce $[BH_2]^+$, $[BH]^+$, and B^+. Loss of hydrogen is also the reason for the misfit between the experimental and simulated spectra in the region of the molecular ion. Superimposed upon the spectrum of the $[B_2H_6]^+$ ion are those of $[B_2H_5]^+$, $[B_2H_4]^+$, etc. Simulation of the envelope of peaks in this region of the mass spectrum is therefore difficult, since the contribution made by each fragmentation product requires additional computation.

(a)   (b)

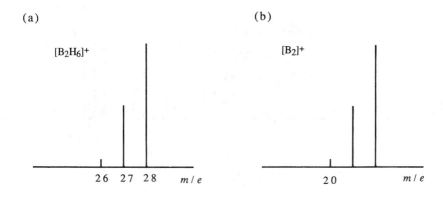

(c)

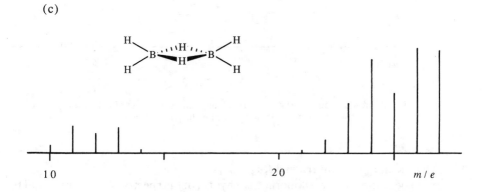

Fig. 3.2.4 — Representations of (a) the calculated molecular ion peaks for $[B_2H_6]^+$, (b) the calculated molecular ion peaks for the B_2^+ ion, and (c) the experimental [5] mass spectrum of B_2H_6.

The description given above typifies the situation for boranes. In a mass spectrometer, hydrogen loss competes with loss of BH_3 or other boron hydride fragments. Consider the experimentally determined mass spectral trace for B_4H_{10} (Fig 3.2.5(a)) [5], which shows loss of BH_3 fragments. Note also that the molecular ion exhibits a peak pattern which fails to mimic that calculated for $[B_4H_{10}]^+$ (Fig. 3.2.5(b)); this again indicates hydrogen loss. An important result is that the observation of four envelopes of peaks may be taken as evidence that the compound possesses a framework based upon four boron atoms.

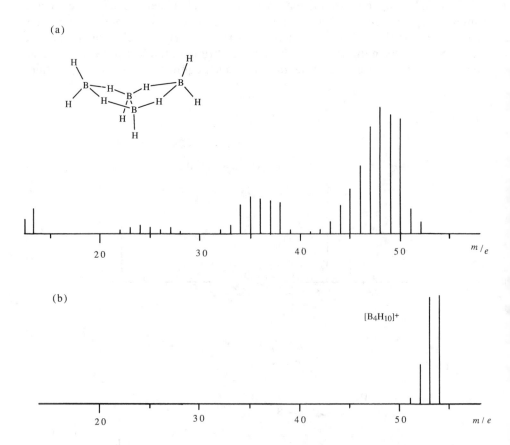

Fig. 3.2.5 — Representations of (a) the experimental [5] mass spectrum of B_4H_{10}, and (b) the calculated molecular ion peaks for the $[B_4H_{10}]^+$ ion.

2. Triruthenium dodecacarbonyl $Ru_3(CO)_{12}$

The mass spectrum of a metalloborane is a hybrid of the spectra of the borane and the metal clusters from which the compound is derived. Since many of the compounds we shall encounter later in this book are derived from transition metal carbonyl complexes, it is instructive to consider the characteristic features of the mass spectrum of a binary metal carbonyl compound.

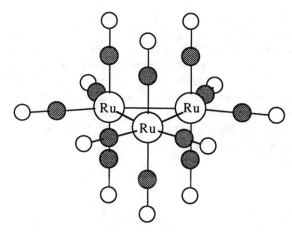

Fig. 3.2.6 — Structure of triruthenium dodecacarbonyl.

In the solid state, triruthenium dodecacarbonyl, $Ru_3(CO)_{12}$, possesses the structure illustrated in Fig. 3.2.6. Each ruthenium atom is bonded in an approximately octahedral environment to four CO ligands and to two other ruthenium atoms. The experimentally determined mass spectrum of $Ru_3(CO)_{12}$ is represented in Fig. 3.2.7. The appearance of thirteen envelopes of peaks, each envelope being separated from the next by 28 mass units, indicates the loss of twelve carbonyl

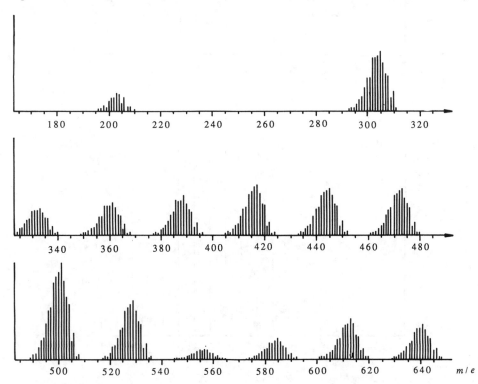

Fig. 3.2.7 — The experimentally determined mass spectrum of $Ru_3(CO)_{12}$.

ligands, and thereby confirms the presence of at least twelve carbonyl ligands in ruthenium carbonyl. This fragmentation pathway is characteristic of transition metal carbonyl complexes, and is often used to confirm the composition of such a compound. Generally, ligand loss predominates in the mass spectral decay pattern of a transition metal cluster leaving the metal core as a detectable fragment ion. The persistence of the triruthenium unit is apparent in Fig. 3.2.7; there is an acceptable match between the experimental mass pattern around $m/e = 304$ a.m.u. in Fig. 3.2.7 and the calculated isotopic distribution for the bare $[Ru_3]^+$ ion illustrated in Fig. 3.2.8(b). The lower intensity peaks in Fig. 3.2.7 centred around m/e 202 are due to the $[Ru_2]^+$ ion.

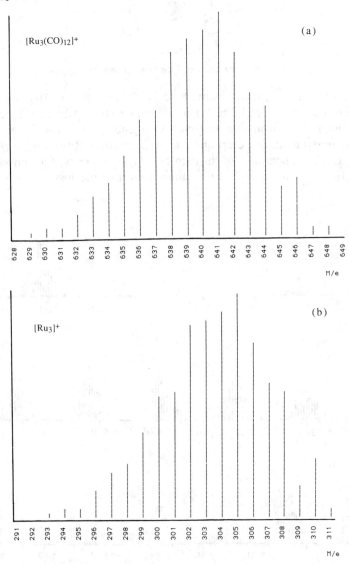

Fig. 3.2.8 — Calculated mass spectra for the molecular ions (a) $[Ru_3(CO)_{12}]^+$ and (b) $[Ru_3]^+$.

Calculated spectra for the $[Ru_3(CO)_{12}]^+$ and the $[Ru_3]^+$ ions are drawn in Figs 3.2.8(a) and 3.2.8(b) respectively. There is excellent agreement between the experimental (Fig. 3.2.7) and simulated (Fig. 3.2.8(a)) isotopic distribution for the molecular ion of $Ru_3(CO)_{12}$. A simulated mass spectrum should coincide with the experimental result in all cases where fragmentation produces ions of well separated masses. In general, the mass spectra of binary transition metal carbonyl complexes exhibit no overlapping envelopes of peaks (compare this situation to the more complicated picture observed for the boranes!). Finally, note the peak of highest m/e in Fig. 3.2.8(a). A comparison of Figs 3.2.8(a) and Fig. 3.2.8(b) illustrates the extent to which the carbon and oxygen nuclei influence the isotopic distribution for the molecular ion, $[Ru_3(CO)_{12}]^+$. The ruthenium isotopic distribution is certainly dominant. Oxygen is essentially monotopic ($^{16}O = 99.759\%$, $^{17}O = 0.037\%$, $^{18}O = 0.204\%$), and thus, in going from $[Ru_3]^+$ to the hypothetical $[Ru_3O_{12}]^+$ ion, the isotopic pattern will remain quite similar. However, after adding in the carbon atoms, (^{12}C and ^{13}C), a significant change appears in the mass spectral pattern; the abundance of $[Ru_3(^{12}C)_{11}(^{13}C)O_{12}]^+ > [Ru_3(^{12}C)_{10}(^{13}C)_2O_{12}]^+ > [Ru_3(^{12}C)_9(^{13}C)_3O_{12}]^+$ etc.

3.3   INFRARED AND RAMAN SPECTROSCOPY

Infrared and Raman spectroscopy are both concerned with vibrational and rotational motions within molecules. For a mode to be *infrared active*, it must give rise to a *change in the electric dipole moment* of the molecule. For a mode to be *Raman active*, it must *change the polarizability* of the molecule. Since the polarizability of a bond is a measure of the electron distribution within the bond, a symmetrical stretch of a symmetrical molecule such as N_2 (Fig. 3.3.1) will be Raman active. On the other hand, the N_2 molecule is non-polar, and a symmetrical stretch of the bond does not alter this property. Hence, this vibrational mode is infrared inactive. (Note that an asymmetric stretch of the N_2 molecule is actually equivalent to a translational degree of freedom.)

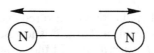

Fig. 3.3.1 — Vibrational mode of the N_2 molecule — the symmetrical stretch.

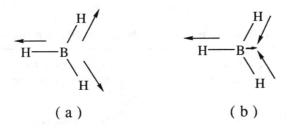

(a)   (b)

Fig. 3.3.2 — Vibrational modes of the BH_3 molecule (a) symmetrical stretch, and (b) asymmetrical stretch.

Consider a molecule of BH_3. Being planar, the molecule possesses no net dipole, although each separate bond is polar. The symmetric stretching mode for BH_3 (Fig. 3.3.2(a)) generates no molecular dipole, and so this mode will be infrared inactive. The vibration is, however, Raman active. As the BH_3 molecule is subjected to the asymmetric stretching mode illustrated in Fig. 3.3.2(b), a molecular dipole *is* generated, and hence, this vibrational mode will be infrared active. The change in polarizability during this vibrational mode is small, and so only a weak Raman absorption is observed.

We should note two important generalizations:

(1) In general, symmetric vibrations give rise to intense Raman absorptions, while asymmetric modes give rise to weak Raman spectral lines.
(2) For a molecule which possesses a centre of symmetry, and **only** for such a molecule, infrared active vibrations are Raman inactive, and vice versa. This is the *rule of mutual exclusion.*

This latter phenomenon allows us to distinguish between isomers of a compound if one isomer is centrosymmetric (i.e. possesses a centre of inversion), and the other is not. An example is shown in Fig. 3.3.3. The *trans*-isomer of $B_2H_4Me_2$ possesses a centre of inversion and will therefore show a diagnostic pattern of infrared active–Raman inactive absorptions.

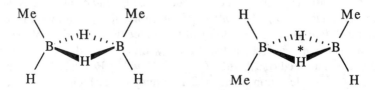

Fig. 3.3.3—Structures of *cis*- and *trans*-isomers of $B_2H_4Me_2$ showing the centre of inversion (*) in the *trans*-isomer.

Perhaps the most useful application of infrared and Raman spectroscopy to the inorganic chemist is in recognizing a characteristic absorption or set of absorptions. This is often referred to as *fingerprinting*. Of the two techniques, infrared spectroscopy is more universally available, and it is to this that we shall refer below. In Table 3.3.1 are listed some abbreviations which are commonly employed in describing spectral lines.

Table 3.3.1 — Abbreviations used in infrared and Raman spectroscopy

ν	Stretch	s	Symmetric
δ	Deformation	as	Asymmetric
s	Strong		
m	Medium	sh	Shoulder
w	Weak	br	Broad

In the infrared spectra of boranes, the absorptions which prove most diagnostic are the B–H(terminal) and the B–H(bridging) stretching modes. These absorptions usually appear in the regions 2600–2400 cm^{-1} and 1500–1800 cm^{-1} respectively. Deformation modes are characterized by lower frequency absorptions. For example, in B_2H_6, the symmetric BH_2(terminal) stretch gives rise to a strong absorption at 2530 cm^{-1}, while an infrared band at 2590 cm^{-1} is assigned to the asymmetric BH_2 stretch. Deformation of the terminal BH_2 unit gives rise to an absorption at 1150 cm^{-1}. In total, there are eight infrared active modes for diborane. In binary boron hydrides, infrared absorptions are readily observed, but in metalloboranes, absorptions due to metal-attached ligand vibrations (e.g. carbonyl stretches) usually predominate in the infrared spectrum with the result that bands assigned to $v_{(B-H)}$ are relatively very weak and may, in fact, not be resolved at all.

3.4 MULTINUCLEAR MAGNETIC RESONANCE SPECTROSCOPY

Nuclei such as ^1H, ^{13}C, ^{19}F, and ^{31}P possess a nuclear spin (I) of $\frac{1}{2}$. This corresponds to nuclear angular momentum, and interaction with an external magnetic field leads to the phenomenon known as *nuclear magnetic resonance* (NMR). In a magnetic field, the nucleus will be aligned either with (low-energy state), or opposed to (high-energy state), the external field. The two spin states are referred to as being parallel and antiparallel, and by the designations $+\frac{1}{2}$ and $-\frac{1}{2}$ respectively. A transition from the lower to upper energy state may be induced by the absorption of radio frequency (RF) radiation. A plot of absorption versus frequency is referred to as an NMR spectrum. Some nuclei are inherently insensitive, and the intensity of absorption is very low. However, this can be increased by recording the spectrum many times and summing the resultant traces. In order to avoid the time-consuming process of repetitive scanning over a range of frequencies, modern techniques rely upon the simultaneous excitation of all nuclei within a given region by the application of an RF pulse. This results in a similar distribution of spin states as that achieved in the direct scanning of frequencies described above. However, instead of recording *absorption*, we measure *emission* of energy. Thus, we obtain a plot of emission versus time (termed the free induction decay, FID), which is dependent upon the time taken for return to the equilibrium spin state. This decay depends upon various mechanisms for *relaxation* (see below). The FID contains identical information to the absorption versus energy plot described above, and the two are related by the mathematical operation of *Fourier transform* (FT), which converts the time to a frequency domain. This method of collecting and treating the data is called Fourier transform spectroscopy; its advantage lies in allowing rapid accumulation of data. By applying a number of sequential RF pulses, a series of FIDs may be collected and summed together; the result is an enhancement of the spectral signal-to-noise ratio. The development of the FT technique has not only increased the usefulness of NMR spectroscopy for common nuclei such as ^1H, ^{13}C, ^{19}F, and ^{31}P, but has also allowed extension of the technique to the study of so-called 'exotic' nuclei, among which is the ^{11}B nucleus. Having chosen the nucleus to be observed, the spectrometer is tuned to the apropriate resonance frequency (Table 3.4.1), a procedure which is directly analogous to tuning a radio: once a 'station' has been selected, only nuclei 'broadcasting' at the selected frequency will be observed.

An isotope must meet certain criteria before it is considered to be an accessible NMR nucleus. It goes without saying that the nucleus must possess a spin ($I > 0$). In addition, the nucleus should preferably be present in a significant natural abundance. This is not an essential criterion, since a compound may be enriched with the nucleus to be observed, and deuterium labelling or enrichment with the ^{13}C isotope is quite commonly encountered. The properties of some representative nuclei are given in Table 3.4.1. (A comprehensive survey of such data may be found in the text by Brevard and Granger [6].)

Table 3.4.1 — Properties of some representative NMR active nuclei

Nucleus	Natural abundance/ %	Nuclear spin	Observe frequency/MHz referred to ^1H at 100 MHz[a]	Chemical shift reference[a]
^1H	>99.9	1/2	100	Me_4Si
^2H	0.02	1	15.35	Me_4Si
^{10}B	19.6	3	10.74	$F_3B.OEt_2$
^{11}B	80.4	3/2	32.07	$F_3B.OEt_2$
^{13}C	1.1	1/2	25.00	Me_4Si
^{19}F	100	1/2	94.0	$CFCl_3$
^{23}Na	100	3/2	26.45	$NaCl(aq)$
^{27}Al	100	5/2	26.06	$[Al(H_2O)_6]^{3+}$
^{29}Si	4.7	1/2	19.87	Me_4Si
^{31}P	100	1/2	40.48	$H_3PO_4(85\% \text{ aq})$
^{69}Ga	60.4	3/2	24.00	$[Ga(H_2O)_6]^{3+}$
^{71}Ga	39.6	3/2	30.50	$[Ga(H_2O)_6]^{3+}$
^{103}Rh	100	1/2	3.19	Rh (metal)
^{107}Ag	51.8	1/2	4.05	$Ag^+(aq)$
^{109}Ag	48.2	1/2	4.65	$Ag^+(aq)$
^{117}Sn	7.6	1/2	35.63	Me_4Sn
^{119}Sn	8.6	1/2	37.29	Me_4Sn
^{195}Pt	33.8	1/2	21.46	$Na_2[PtCl_6]$

[a] The operating frequency of an instrument is defined by the field of the magnet, and is designated by the frequency at which the ^1H nuclei of Me_4Si resonate.
[b] Although references for most nuclei are standard, changes may have been made with time. For instance, $B(OMe)_3$ has also been used as a standard reference for ^{11}B NMR spectra. Thus, the reader should beware, and should pay attention when referring to the chemical literature.

A further requirement for a potential NMR nucleus is that it should exhibit a relatively short *spin-lattice relaxation time* (T_1). This is related to the time taken to reestablish the ground state populations of the nuclear spin states. The excess energy gained on excitation is lost to the surroundings. Values of T_1 are dependent not only upon the nucleus itself, but also on the environment of the nucleus. For example, a proton attached to a carbon atom usually relaxes at a very different rate from a proton attached to one or more metal atoms. Similarly, values of T_1 for ^{31}P nuclei vary over a range of 0.1–55 s. (A second relaxation process involves spin–spin relaxation and is characterized by a time T_2.) Clearly, it is not particularly convenient to run an NMR experiment if a nucleus requires a long delay period between successive nuclear excitations. The observation of ^6Li versus ^7Li NMR spectra provides a useful example: T_1 values for ^{6}Li nuclei typically lie in the range 10–80 s, whereas T_1 for ^{7}Li is generally <3 s. Since ^{7}Li occurs naturally with an abundance of 93% anyway, this nucleus would most certainly be observed in preference to ^{6}Li.

We see from Table 3.4.1 that not all nuclei possess a spin of $\frac{1}{2}$. Indeed, the nucleus most relevant to boron chemists, ^{11}B, has $I=\frac{3}{2}$. Any nucleus with $I\geqslant1$ also exhibits an electric quadrupole moment. This arises from a non-spherical charge distribution of the nucleus. The quadrupole moment provides an efficient mechanism for relaxation, and thus, values of T_1 for nuclei such as ^{11}B are short. Thus, rapid accumulation of data is possible, and this often compensates for the fact that the sensitivity of the nucleus may be low. A less desirable side effect of a quadrupole moment is that it tends to give rise to broad signals in an NMR spectrum.

Having chosen a nucleus to observe, and having recorded a spectrum, what information may we obtain? A chemist is usually interested in three major features of an NMR spectrum:

(1) chemical shift values
(2) relative areas of signals
(3) nuclear spin–spin coupling patterns.

Chemical shift values, δ, are recorded with respect to a standard reference (Table 3.4.1), and are independent of the magnetic field. Thus, for ^{11}B NMR spectra, the chemical shift of a signal (or resonance) is given relative to $F_3B.OEt_2$ at $\delta=0$. Although there have been many changes over past years, it is now commonly accepted that signals appearing at *low field* (downfield) have *positive* chemical shifts, and those *upfield* from the reference are assigned *negative* δ values. Chemical shift values are often solvent dependent, and are certainly highly dependent upon the environment of the nucleus. The latter is well exemplified by ^{1}H NMR spectra. Protons which are in conventional organic environments typically show resonances in the region of δ 0 to $+10$ [4]. However, inorganic protons may occur over a larger chemical shift range. Note that the ranges illustrated in Fig. 3.4.1 are approximate; for example, in the case of a metal hydride, the value of δ (^{1}H NMR spectrum) varies with the metal. An illustration is given in Fig. 3.4.2 in which we compare the chemical shift values for the metal- and boron-associated hydrogen atoms in the metalloborane clusters, $HM_4(CO)_{12}BH_2$ (M=Fe, Ru).

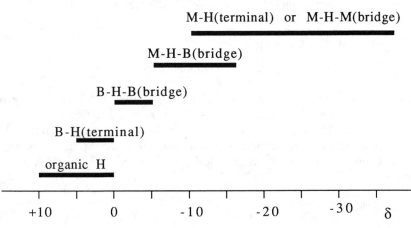

Fig. 3.4.1 — Representative chemical shift ranges in a ^{1}H NMR spectrum.

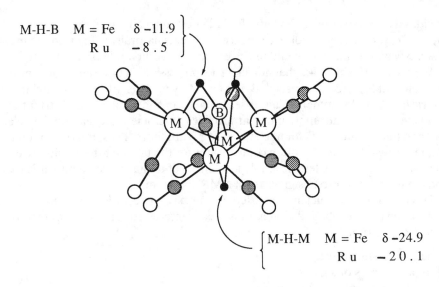

M-H-B   M = Fe δ −11.9
         R u   − 8 . 5

M-H-M   M = Fe δ −24.9
         R u   − 2 0 . 1

Fig. 3.4.2 — The structure of $HM_4(CO)_{12}BH_2$ (M = Fe, Ru) and the chemical shift values in the ^1H NMR spectra showing the effect of changing the metal atom.

The area under an NMR spectral peak (the peak *integral*) provides an indication of the number of nuclei in a given environment. By measuring the relative areas of several peaks in a spectrum (*integration*), we should be able to write down the ratio of nuclei in different bonding modes. However, owing to differences in the relaxation mechanisms of nuclei in dissimilar environments, one should always be cautious in comparing peak integrals. Once again, ^1H NMR spectroscopy provides an excellent example; protons attached to metal atoms typically exhibit extremely low integrals when compared to carbon-bonded protons. Thus, interpreting relative integral values in terms of the ratio of nuclei present is only valid if we restrict our comparisons to nuclei in similar bonding modes.

The interaction, or *coupling*, of nuclear spins is an important aspect of NMR spectroscopy. Coupling patterns depend upon the numbers and actual spins of the nuclei which are interacting. The *multiplicity* of an NMR signal is the number of lines making up the signal and is given by the equation

Multiplicity = $(2nI + 1)$

n = number of coupled *equivalent* nuclei, each of spin I.

The magnitude of a spin–spin interaction between two nuclei X and Y is given by the *coupling constant*, J_{XY}, and is expressed as a frequency difference measured in Hz. Note that the *difference in Hz* will be independent of the magnetic field, whereas the *difference in δ* will not. For boranes it is useful to remember that the coupling of a boron nucleus to terminal hydrogen atoms is often the dominant feature of the spectrum, since

$J_{BH(terminal)} > J_{BH(bridging)}$

In addition, for the same pair of atoms, one of which is a boron atom and the other a spin active nucleus, X:

$$J_{^{11}BX} > J_{^{10}BX}$$

Let us illustrate the use of coupling patterns by considering the molecule B_2H_6. There are three alternative spin active nuclei which we might observe: ^1H ($I = \frac{1}{2}$), ^{10}B ($I = 3$), or ^{11}B ($I = \frac{3}{2}$). Of the two boron nuclei, it is more convenient to observe ^{11}B, for reasons of more favourable natural abundance and nuclear spin (see Table 3.4.1). The two boron atoms in B_2H_6 are equivalent and each one is attached to two equivalent terminal and to two equivalent bridging hydrogen atoms (Fig. 3.4.3). Consider, first, only the terminal atoms. Each ^{11}B nucleus may interact, with equal probability, with the two terminal ^1H nuclei, the spin state of each being either $+\frac{1}{2}$ or $-\frac{1}{2}$. The resultant coupling pattern, a 1:2:1 *triplet*, is constructed in Fig. 3.4.3 by considering separately the coupling of each ^{11}B nucleus to each ^1H nucleus. The multiplicity of the signal may be confirmed as follows:

$$\text{Multiplicity} = 2nI + 1 = 2(2)(\tfrac{1}{2}) + 1 = 3$$

Each line of the ^{11}B spectral resonance is then further affected by the bridging hydrogen atoms. Again, the presence of two equivalent nuclei with $I = \frac{1}{2}$ gives rise to a triplet as shown in Fig. 3.4.3. The final ^{11}B NMR signal is thus a *triplet of triplets*. Note, however, that, depending upon the relative magnitudes of $J_{^{11}BH(terminal)}$ and $J_{^{11}BH(bridge)}$, the triplets may overlap.

Let us now turn to the ^1H NMR spectrum of B_2H_6. There will be one resonance due to the two equivalent bridging hydrogen atoms, and this is at higher field with respect to a signal due to the four equivalent terminal hydrogen atoms (see Fig. 3.4.1). Each terminal hydrogen nucleus will interact with one boron atom. For $\approx 80\%$ of the protons, this will be an ^{11}B nucleus, and for $\approx 20\%$, it will be a ^{10}B nucleus. Thus the major features in the spectrum arise from $^1H-^{11}B$ spin–spin coupling. Since for the ^{11}B nucleus, $I = \frac{3}{2}$, there are four possible spin states (i.e. $+\frac{3}{2}$, $+\frac{1}{2}$, $-\frac{1}{2}$, $-\frac{3}{2}$) with which a neighbouring nucleus may interact. Coupling with each state occurs with an equal probability, and thus the ^1H resonance for the terminal hydrogen atoms splits into a 1:1:1:1 four-line multiplet. The multiplicity is confirmed by

$$\text{Multiplicity} = 2nI + 1 = 2(1)(\tfrac{3}{2}) + 1 = 4$$

Finally, each bridging hydrogen nucleus interacts with two ^{11}B nuclei. The multiplicity of this ^1H NMR signal is given by

$$\text{Multiplicity} = 2nI + 1 = 2(2)(\tfrac{3}{2}) + 1 = 7$$

The predicted ^1H NMR spectrum is represented in Fig. 3.4.3. Note that the final 1:2:3:4:3:2:1 seven-line multiplet is constructed by considering the coupling of each pair of nuclei in turn, and by remembering that each coupling constant is identical.

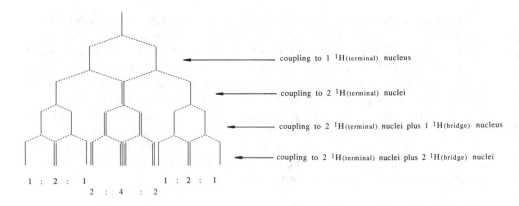

coupling to 1 ^{1}H$_{(terminal)}$ nucleus

coupling to 2 ^{1}H$_{(terminal)}$ nuclei

coupling to 2 ^{1}H$_{(terminal)}$ nuclei plus 1 ^{1}H$_{(bridge)}$ nucleus

coupling to 2 ^{1}H$_{(terminal)}$ nuclei plus 2 ^{1}H$_{(bridge)}$ nuclei

^{11}B nmr spectrum for B$_2$H$_6$: Triplet of triplets:

^{1}H nmr spectrum for the terminal H atoms

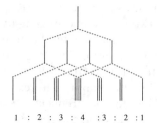

^{1}H nmr spectrum for the bridging H atoms

Fig. 3.4.3 — The structure of B$_2$H$_6$ and the simulated ^{11}B and ^{1}H NMR spectra.

The spectrum described above allows us to appreciate the origins of line intensities. Relative intensities reflect the relative probabilities of the occurrence of particular combinations of nuclear spin states. Thus, although coupling either to one nucleus with $I = \frac{3}{2}$ or to three equivalent nuclei each of $I = \frac{1}{2}$ gives a four-line multiplet, the former gives rise to a 1:1:1:1 pattern whilst the latter results in a 1:3:3:1 quartet. In the case of **equivalent** nuclei with spin $\frac{1}{2}$, the NMR signal pattern will always reflect a binomial distribution of spin combinations, represented by Pascal's triangle shown in Fig. 3.4.4.

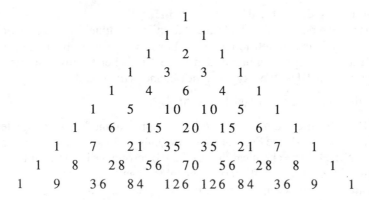

```
                        1
                    1       1
                1       2       1
            1       3       3       1
        1       4       6       4       1
    1       5      10      10       5       1
  1     6      15      20      15       6       1
1     7      21      35      35      21       7       1
  1     8      28      56      70      56      28       8       1
1     9      36      84     126     126      84      36       9       1
```

Fig. 3.4.4 — Peak intensities for a nuclear spin coupling to n **equivalent** nuclei each with $I=\frac{1}{2}$.

We shall now consider several case histories, each illustrated with experimental NMR spectra. The spectra have been recorded on NMR spectrometers operating at proton frequencies of 250, 300, or 400 MHz. The ^{11}B and ^{31}P observed frequencies also vary according to the instrument, and are defined in each figure caption below.

1.   Borohydride anion [BH$_4$]$^-$

The borohydride anion, [BH$_4$]$^-$, is isoelectronic and isostructural with CH$_4$. The ^{11}B NMR spectrum of an aqueous solution of sodium borohydride is illustrated in Fig. 3.4.5. The 1:4:6:4:1 quintet is indicative of the ^{11}B nucleus coupling to four equivalent ^{1}H nuclei. This is consistent with the tetrahedral structure of the [BH$_4$]$^-$ ion. The coupling constant, J_{BH}, of 82 Hz is relatively small for terminal hydrogen atom coupling.

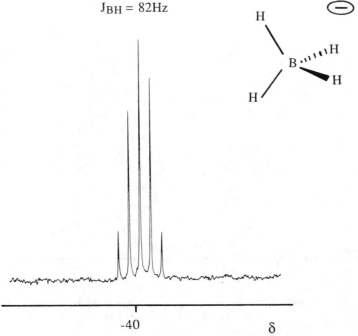

Fig. 3.4.5 — The 128 MHz ^{11}B NMR spectrum for Na[BH$_4$] in D$_2$O solution.

2.   Lewis base adducts, thf.BH$_3$ and Et$_3$N.BH$_3$

In Chapter 2, we discussed the relative stabilities of adducts formed between Lewis bases and BH$_3$. The competition between Lewis bases is readily monitored by using ^{11}B NMR spectroscopy, as illustrated in Fig. 3.4.6. When triethylamine is added to a thf solution of thf.BH$_3$, the oxygen donor ligand is displaced:

$$thf.BH_3 + Et_3N \rightarrow Et_3N.BH_3 + thf$$

If the ^{11}B NMR spectrum is recorded after approximately half equivalent of the amine has been added (see Fig. 3.4.6), two signals are observed, one for thf.BH$_3$ and one for Et$_3$N.BH$_3$. The chemical shift of the ^{11}B resonance is significantly affected by the nature of the base. Addition of a phosphine ligand in place of the amine would result in an even greater upfield chemical shift; see Fig. 3.4.7.

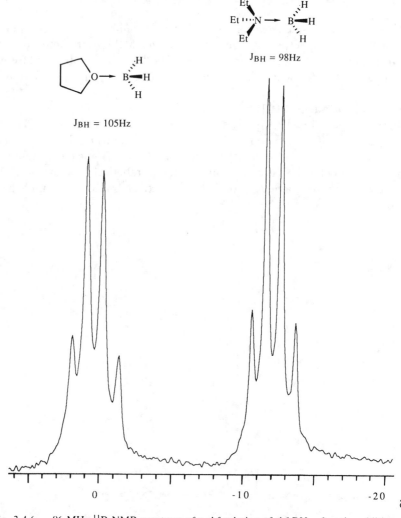

Fig. 3.4.6 — 96 MHz ^{11}B NMR spectrum of a thf solution of thf.BH$_3$ after the addition of approximately half equivalent of Et$_3$N.

Each resonance shown in Fig. 3.4.6 is a 1:3:3:1 quartet. This line multiplicity arises from the coupling of the ^{11}B nucleus to three equivalent ^{1}H nuclei:

$$\text{Multiplicity} = 2nI + 1 = 2(3)(\tfrac{1}{2}) + 1 = 4$$

Note that any coupling of the ^{11}B nucleus to ^{17}O (0.037%, $I=\tfrac{5}{2}$) or ^{15}N (0.37%, $I=\tfrac{1}{2}$) may be ignored on the grounds of the low isotopic abundance of these nuclei. However, the ^{11}B nucleus *will* couple to ^{14}N (99.63%, $I=1$) but the coupling is not resolved.

3.   Lewis base adduct, PhMe$_2$P.BH$_3$

If a compound contains several spin active nuclei, it is often helpful to record the NMR spectrum for *each* nucleus in order to confirm conclusions drawn from any one spectrum concerning the structure of the compound. The borane adduct, PhMe$_2$P.BH$_3$, provides an appropriate example. There are five spin active nuclei present, viz. ^{1}H, ^{13}C, ^{11}B, ^{10}B, and ^{31}P. The complex may be formed by ligand displacement:

$$\text{thf.BH}_3 + \text{PhMe}_2\text{P} \rightarrow \text{PhMe}_2\text{P.BH}_3 + \text{thf}$$

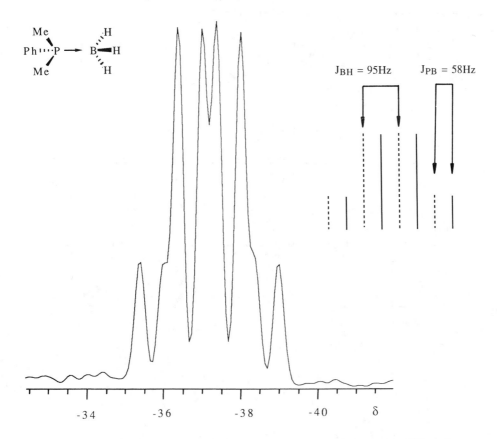

Fig. 3.4.7 — 96 MHz ^{11}B NMR spectrum of a thf solution of thf.BH$_3$ after the addition of an excess of PhMe$_2$P.

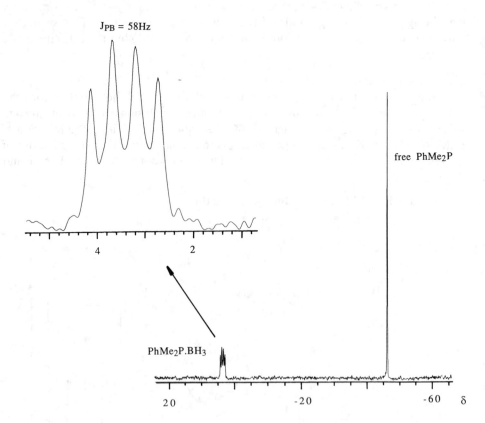

Fig. 3.4.8 — 121.5 MHz ^{31}P NMR spectrum of a thf solution of thf.BH$_3$ after the addition of an excess of PhMe$_2$P.

Figs 3.4.7 and 3.4.8 illustrate, respectively, the ^{11}B and ^{31}P NMR spectra after an excess of the phosphine has been added to a solution of thf.BH_3. The spectra need a little thought prior to interpretation! Consider first the ^{11}B NMR spectroscopic data. The ^{11}B nucleus couples to the three equivalent ^{1}H nuclei, thereby giving rise to a 1:3:3:1 quartet (see Fig. 3.4.6). There is then additional coupling to the ^{31}P nucleus ($I=\frac{1}{2}$). The net result is a *doublet of quartets*, and this coupling pattern is quite clearly resolved in Fig. 3.4.7. In order to confirm the assignment of the ^{11}B signal, we should now look at the ^{31}P NMR spectrum which is shown in Fig. 3.4.8. Since an excess of dimethylphenylphosphine was initially added to the solution, two ^{31}P NMR resonances are expected, one for the adduct, $PhMe_2P.BH_3$, and one for free $PhMe_2P$. The signal at δ −45 in Fig. 3.4.8 corresponds to the free phosphine. The four-line multiplet centred at δ +3.4 (inset in Fig. 3.4.8) exhibits relative line intensities which are characteristic when coupling to a boron nucleus is involved. Why is a clean 1:1:1:1 multiplet (corresponding to ^{31}P–^{11}B coupling) not observed? The answer lies in the fact that only ≈80% of the ^{31}P nuclei are coupled to ^{11}B. The remaining 20% couple to the second spin active nucleus, ^{10}B ($I=3$). The multiplicity of this second signal is given by

$$\text{Multiplicity} = 2nI + 1 = 2(1)(3) + 1 = 7$$

Since the value of $J_{^{31}P-^{10}B} < J_{^{31}P-^{11}B}$, the seven-line 1:1:1:1:1:1:1 multiplet due to the ^{31}P–^{10}B interaction underlies only the central portion of the four-line multiplet due to the ^{31}P–^{11}B coupling. Thus, the 1:1:1:1 multiplet swamps the seven-line signal, and we simply observe an enhancement of the central two lines of the former. Relaxation effects also contribute to the line shape.

In conclusion, we see that the ^{31}P NMR spectroscopic data confirm the presence of a direct P–B interaction, and aid the interpretation of the composite ^{11}B NMR spectrum.

4. Octahydrotriborate anion, $[B_3H_8]^-$

The solid-state structure of $[PhCH_2NMe_3][B_3H_8]$ has been determined by X-ray crystallography, and the geometry of the anion is represented in Fig. 3.4.9. (The bonding in this anion will be discussed in section 6.3.) The anion possesses a plane of symmetry and there are two boron (in a ratio 1:2) and three proton environments (in a ratio 2:4:2). Hence, in the ^{11}B NMR spectrum we might expect to see a triplet of triplets for the unique ^{11}B nucleus and a doublet of triplets due to the two equivalent boron atoms. In the ^{1}H NMR spectrum of $[B_3H_8]^-$, we would predict two 1:1:1:1 multiplets, one for each set of equivalent terminal hydrogen atoms, and one 1:2:3:4:3:2:1 line resonance for the bridging protons. Compared to these predicted spectra, the observed room temperature ^{11}B and ^{1}H NMR spectra (Figs 3.4.9 and 3.4.10 respectively) look significantly different! The reason for this is that, in solution, the hydrogen atoms in the octahydrotriborate(1−) anion are *fluxional*. Each H atom spends a finite time in *each* of the available terminal and bridging locations, including the unique bridging site which, in the solid-state structure, appears as an unbridged B–B bond. A mechanism for hydrogen atom exchange is represented in Fig. 3.4.11. Provided that the rate of exchange is *faster* than the NMR timescale, all eight hydrogen atoms will appear to be equivalent. In addition, the three boron atoms will appear equivalent, and the observed coupling constant, $J_{^{11}B-^1H(observed)}$, will be a weighted average of the terminal and bridging coupling constants.

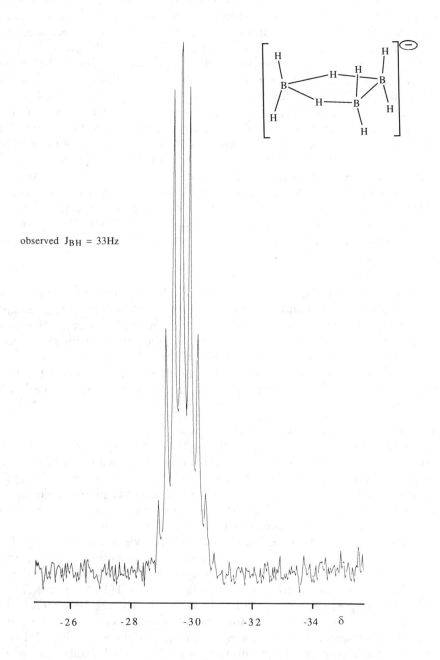

observed J_{BH} = 33Hz

Fig. 3.4.9 — Structure of the $[B_3H_8]^-$ anion and room temperature 128 MHz ^{11}B NMR spectrum of $[Me_4N][B_3H_8]$ in d_7-dimethylformamide solution.

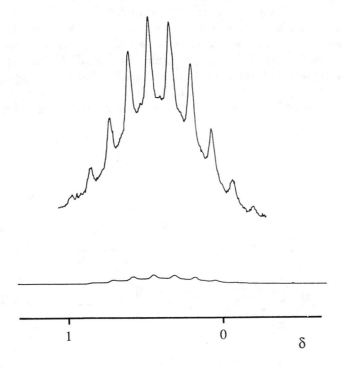

observed J_{BH} = 33Hz

Fig. 3.4.10 — Room temperature 400 MHz ^{1}H NMR spectrum of [Me$_4$N][B$_3$H$_8$] in d_7-dimethylformamide solution.

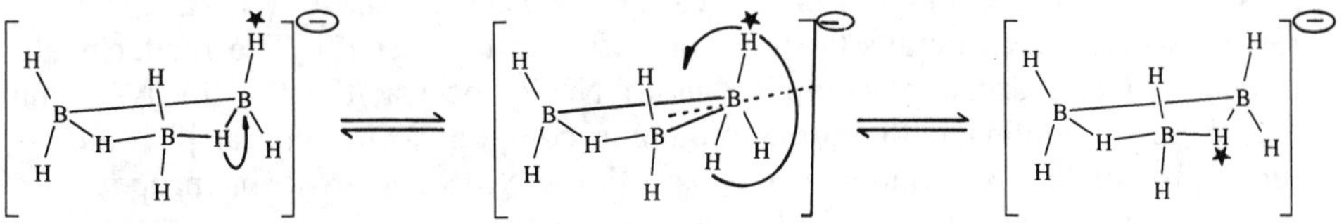

Fig. 3.4.11 — Fluxional process interconverting terminal and bridging hydrogen atoms in the [B$_3$H$_8$]$^-$ anion.

The experimental NMR spectra in Figs 3.4.9 and 3.4.10 are interpreted as follows. Each ^{11}B nucleus couples to *all eight* of the ^{1}H nuclei; a nine-line multiplet (nonet), with intensity ratios $1:8:28:56:70:56:28:8:1$ (Fig. 3.4.4), is therefore predicted. Compare this with the observed spectrum shown in Fig. 3.4.9. Only seven lines have been completely resolved. Even so, the signal cannot be mistaken for a binomial seven-line multiplet since this would require line intensities of $1:6:15:20:15:6:1$. The ^{1}H NMR spectrum of the $[B_3H_8]^-$ ion consists of an extremely broad signal, shown in the lower trace in Fig. 3.4.10. An expansion of the signal reveals a ten-line multiplet, consistent with the coupling of each ^{1}H nucleus with three equivalent ^{11}B nuclei. The spectrum is simulated in Fig. 3.4.12.

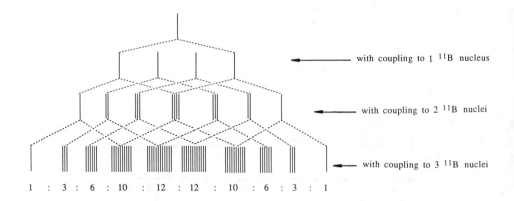

with coupling to 1 ^{11}B nucleus

with coupling to 2 ^{11}B nuclei

with coupling to 3 ^{11}B nuclei

1 : 3 : 6 : 10 : 12 : 12 : 10 : 6 : 3 : 1

Fig. 3.4.12 — Simulation of the ^{1}H NMR spectrum for the fluxional $[B_3H_8]^-$ ion.

The phenomenon of fluxional behaviour is common in borane and metalloborane compounds, particularly when an asymmetrical arrangement of hydrogen atoms is observed in the solid state. Activation energies for fluxional processes obviously vary with the system involved. The $[B_3H_8]^-$ anion provides one example of an extremely facile interconversion of terminal and bridging hydrogen atoms, and the process persists even at low temperatures. Beall and Bushweller have written a valuable review article which details fluxional processes in boranes [7]. The reader is also directed to two excellent compilations of NMR spectra, the first by Nöth and Wrackmeyer dealing with boron-containing compounds in general [8], and the second by Siedle and Todd which deals with the NMR spectroscopic properties of boranes, carboranes and metalloboranes [9].

3.5   DIFFRACTION METHODS

Spectroscopic studies provide valuable information about the structure of a compound, and indeed, a structure may well be unambiguously assigned upon this basis. Confirmation, however, is usually sought by using a diffraction technique, and a review [10], (comprehensive up to 1987) of the molecular structures of neutral boranes and carboranes will prove useful to readers of this book. The results of a diffraction study allow the geometrical coordinates for the constituent atoms in a compound to be determined. A single crystal X-ray diffraction study is probably the most commonly employed method. X-rays are diffracted by the electrons of an atom. For most atoms, the nucleus may be assumed to lie at the centre of the core electron density; however, in the case of a covalently bound hydrogen atom, the only electron density is that associated with an X–H bond. Consequently, location of hydrogen atoms is not always a trivial matter, particularly in the presence of transition metal atoms. Even when direct location is possible, the X–H internuclear distance will be *underestimated* by X-ray diffraction. For some borane molecules, structural determination in the gas phase by electron diffraction may be a useful alternative to X-ray diffraction. A pertinent example is B_6H_{12}, the structure of which (see section 4.6) has recently been determined by electron diffraction, after attempts to crystallize the borane had proved unsuccessful [11]. The fact that electron diffraction is a gas phase technique means that the structural study is made on a molecule which is undergoing rotational and vibrational motions. Lowering the temperature may reduce this tendency. Bonding and nonbonding interatomic distances are determined directly, and the data are then used to calculate bond angles. By using neutron diffraction, all atoms, including hydrogen, can be accurately located. However, the restricted availability of this technique and the fact that relatively large crystals are required make it less attractive than either electron or X-ray diffraction.

Electron and X-ray diffraction techniques can provide *different* values for a given interatomic distance depending upon the phase of material being investigated. This is exemplified by bond parameters for B_2H_6, the structure of which has been determined both in the gas phase by electron diffraction, and in the solid state by X-ray analysis. Experimental data are provided in Table 3.5.1, and the gas phase results are compared with those obtained by using microwave spectroscopy [1].

The results of X-ray diffraction studies provide insight into the ways in which molecules pack together in the solid state, and the way in which solvent molecules are incorporated into the crystal lattice. Examination of a molecular packing diagram may provide explanations for physical properties of the bulk material. Distortions in molecular geometry may be imposed by crystal packing forces with the energy barrier for deformation being ≈ 12 kJ mol^{-1} at room temperature.

Table 3.5.1 — Structural parameters for B_2H_6

Parameter	X-ray diffraction [12]	Electron diffraction [13]	Microwave spectroscopy [14]
	Bond distance (Å)		
B–B	1.762	1.775	1.7693
B–H(terminal)	1.09	1.196	1.1936
B–H(bridge)	1.24	1.339	1.3273
	Bond angle (deg)		
H(terminal)–B–H(terminal)	121.6	119.0	121.65
H(bridge)–B–H(bridge)	90	97.0	96.41

REFERENCES

[1] *Structural Methods in Inorganic Chemistry*, E. A. V. Ebsworth, D. W. H. Rankin and S. Cradock, Blackwell Scientific Publications, Oxford, 1987.

[2] *Physical Methods in Chemisty*, R. S. Drago, Saunders Co., London, 1977.

[3] *Modern NMR Spectroscopy: a Guide for Chemists*, B. K. Hunter and J. K. M. Saunders, Oxford University Press, Oxford, 1986.

[4] *Spectroscopic Methods in Organic Chemistry*, 4th edn, D. H. Williams and I. Fleming, McGraw-Hill, 1988.

[5] I. Shapiro, C. O. Wilson, J. F. Ditter and W. L. Lehmann, *Adv. Chem. Series*, 1961, **32**, 127.

[6] *Handbook of High Resolution Multinuclear NMR*, C. Brevard and P. Granger, Wiley–Interscience, New York, 1981.

[7] H. Beall and C. H. Bushweller, *Chem. Rev.*, 1973, **73**, 465.

[8] *Nuclear Magnetic Resonance Spectroscopy of Boron Compounds*, H. Nöth and B. Wrackmeyer, Springer-Verlag, Berlin, 1978.

[9] A. R. Siedle and L. J. Todd, *Prog. in N.M.R. Spectrosc.*, 1979, **13**, 87.

[10] R. A. Beaudet, Chapter 20 in *Advances in Boron and the Boranes*, eds J. F. Liebman, A. Greenberg and R. E. Williams, VCH Publishers, Weinheim, 1988.

[11] R. Greatrex, N. N. Greenwood, M. B. Millikan, D. W. H. Rankin and H. E. Robertson, *J. Chem. Soc., Dalton Trans.*, 1988, 2335.

[12] D. S. Jones and W. N. Lipscomb, *J. Chem. Phys.*, 1969, **51**, 3133.

[13] L. S. Bartell and B. L. Carroll, *J. Chem. Phys.*, 1965, **52**, 1135.

[14] J. L. Duncan and J. Harper, *Mol. Phys.*, 1984, **51**, 371.

4

Structures of boranes and hydroborate anions

4.1 INTRODUCTION

In this chapter, we shall survey the structures of the neutral boranes and the hydroborate anions. The discussion includes not only compounds for which structures have been confirmed by either an electron or an X-ray diffraction study, but also boron hydrides for which structural assignments have been made on the basis of spectroscopic data. The borane structures described are grouped according to the number of boron atoms in the compound, and also according to the nature of the cage. Sections 4.2–4.12 survey structures comprising a single unit, while structures involving fused borane units are described together in section 4.13. Methods of rationalizing the structures will be discussed in Chapter 6.

With the exception of B_2H_6 and the $[B_2H_7]^-$ ion, the binary boron hydrides and their anions adopt cage-like structures in which triangular B_3-faces are the predominant feature. Each cage geometry is derived from that of a regular deltahedron (i.e. triangular-faced polyhedron). In the case of some of the larger boranes, e. g. $B_{18}H_{22}$, the cage geometry is derived from two fused deltahedra. The deltahedra to which we shall refer throughout this and subsequent chapters are illustrated in Fig. 4.1.1. Notice that only in the tetrahedron ($n = 4$), the octahedron ($n = 6$), and the icosahedron ($n = 12$) are all the cage vertices equivalent in terms of connectivity.

The connectivity of a boron atom in a cage is defined here as being the number of adjacent boron atoms. For example, in an octahedral cage, each boron atom has a connectivity of 4. All hydrogen atoms are ignored.

Note: A terminal hydrogen atom which is attached to the borane cluster via a localized 2-centre 2-electron bond is termed an *exo*-hydrogen. Hydrogen atoms bound intimately to the cluster (i.e. the bonding involves cluster bonding MOs) are called *endo*-hydrogen atoms. The terms *exo*- and *endo*- also apply to ligands other than hydrogen atoms.

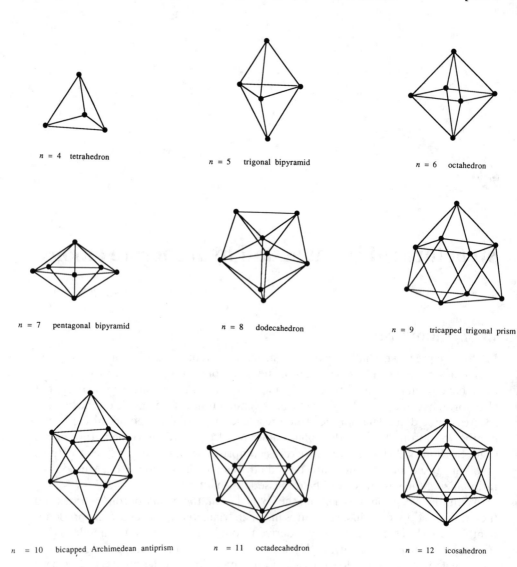

Fig. 4.1.1 — Regular n-vertex deltahedra.

4.2 TWO BORON ATOMS

As mentioned above, diborane(6) [1] and the $[B_2H_7]^-$ anion [2] (Fig. 4.2.1) possess open structures. Among the boranes, B_2H_6 is highly unusual in that it exhibits *two* hydrogen atoms bridging a single B–B bond. On the other hand, the $[B_2H_7]^-$ ion is remarkable because it exhibits an *unsupported* B–H–B bond, i.e. the two borane fragments (BH_3 units) are held together *solely* by a single bridging hydrogen atom.

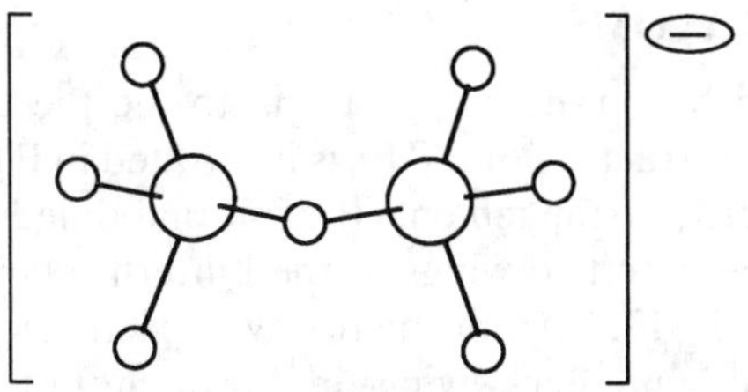

Fig. 4.2.1 — Crystallographically determined structure of the $[B_2H_7]^-$ ion.

The results of X-ray diffraction studies for B_2H_6 and $[B_2H_7]^-$ are listed in Table 4.2.1, and one notable feature is that in the two compounds, the angle subtended at the bridging hydrogen atom is markedly different. In the presence of only one hydrogen atom, the bridge relaxes open with the result that the boron atoms in $[B_2H_7]^-$ are much further apart than in B_2H_6.

Table 4.2.1 — Structural parameters for B_2H_6 and $[B_2H_7]^-$ obtained from X-ray diffraction studies [1,2]

	B_2H_6	$[B_2H_7]^-$
Bond length (Å)		
B–B	1.762	2.107
B–H(terminal)	1.09	1.03
B–H(bridge)	1.24	1.14
Bond angle (deg)		
H(terminal)–B–H(terminal)	121.6	110
B–H(bridge)–B	90	136

4.3   THREE BORON ATOMS

The $[B_3H_8]^-$ ion stands alone as an example of a stable boron hydride comprising three boron atoms. In Chapter 3, we discussed the fluxional properties exhibited by $[B_3H_8]^-$. In the solid state [3], the boron atoms define an isosceles triangle (Fig. 4.3.1), with distances for the unbridged and bridged B–B bonds of 1.804 Å and 1.769 Å (average) respectively. The bridging hydrogen atoms are positioned asymmetrically, being further from the unique boron atom of the B_3-triangle: average B–H(bridge) = 1.37 Å and 1.07 Å.

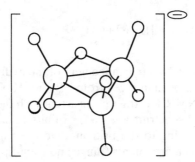

Fig. 4.3.1 — Crystallographically determined structure of the $[B_3H_8]^-$ ion [3].

4.4   FOUR BORON ATOMS

For boranes with ≥ 4 boron atoms, we begin to see the emergence of a cage, or cluster, structure. The structure of B_4H_{10} is illustrated in Fig. 4.4.1 [4,5]. The boron atoms define a 'butterfly' arrangement, the internal dihedral angle of the butterfly (i.e. the angle between the two wings of the butterfly) being 124.5° from electron diffraction results and 117.4° from microwave spectroscopy. Each 'hinge' boron atom is bonded to one, and each 'wing-tip' boron atom to two terminal hydrogen atoms. This distinction is necessary in view of the different connectivity numbers (see section 4.1) of the two types of boron atoms.

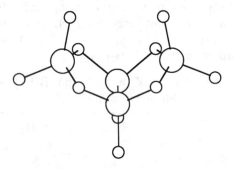

Fig. 4.4.1 — Crystallographically determined structure of B_4H_{10}.

The unbridged B–B bond in B_4H_{10} is shorter (1.750 Å) than the bridged B–B bonds (1.845 Å). This is the reverse of the situation observed in the $[B_3H_8]^-$ ion, and the difference is probably due to the nature of the bridging hydrogen atoms. The interaction of a *hydride* ion, H^-, with a B–B bond has the effect of generating an additional occupied B–H–B bonding molecular orbital. This bonding character *enhances* the B–B interaction and causes bond shortening. On the other hand, interaction of a B–B bonding orbital with a *proton*, H^+, to form a 3-centre 2-electron bond results in a reduction of direct B–B bonding character and, hence the B–B bond lengthens [3].

The acidic nature of the bridging hydrogen atoms in B_4H_{10} is illustrated by the deprotonation reactions:

$$B_4H_{10}+KH \rightarrow K[B_4H_9]+H_2$$

$$B_4H_{10}+LiCH_3 \rightarrow Li[B_4H_9]+CH_4$$

Deprotonation is followed by a rearrangement of the hydrogen atoms; the structure of the $[B_4H_9]^-$ anion has been inferred from ^{11}B and ^{1}H NMR spectroscopic studies [6]. Low-temperature NMR spectra are consistent with the static structure shown in Fig. 4.4.2; the ^{11}B NMR spectrum at $-90°C$ exhibits a doublet (1B) and two triplets (2B:1B); note that coupling to the bridging hydrogen atoms is not resolved. At $+40°C$, only two boron environments (hinge and wing-tip) are observed, consistent with a fluxional process involving the hydrogen atoms.

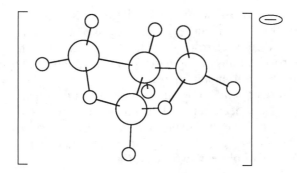

Fig. 4.4.2 — Proposed structure of the $[B_4H_9]^-$ anion.

4.5 FIVE BORON ATOMS

The two pentaboranes which have been crystallographically characterized are B_5H_9 and B_5H_{11}. The structure of pentaborane(9) exhibits a square-based pyramidal B_5-core (Fig. 4.5.1), and the two boron environments are readily distinguished by using ^{11}B NMR spectroscopy. *Apical boron atoms appear at higher field in the ^{11}B NMR spectrum than do basal boron atoms.* Bond distances for B_5H_9, obtained from X-ray and electron diffraction studies, are given in Table 4.5.1; as with B_4H_{10}, the unbridged B–B bonds in B_5H_9 are shorter that the bridged bonds.

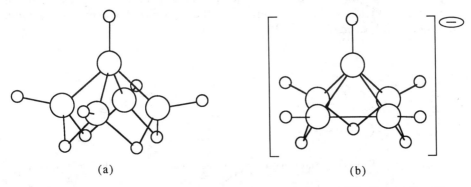

(a)   (b)

Fig. 4.5.1 (a) Structure of B_5H_9 confirmed by X-ray and electron diffraction studies; (b) proposed structure of $[B_5H_8]^-$.

Table 4.5.1 — Structural parameters for B_5H_9 obtained from X-ray [7] and electron diffraction [8] studies

	X-ray diffraction	Electron diffraction
Bond length (Å)		
B(apical)–B(basal)	1.66	1.694
B(basal)–B(basal)	1.72	1.811
B–H(terminal) (av.)	1.20	1.210
B–H(bridge)	1.35	1.374

Deprotonation of B_5H_9 to generate $[B_5H_8]^-$ is achieved by using alkali metal hydride or methyl lithium, and proceeds by loss of a proton from a B–H–B bridging site [9]. The NMR spectroscopic data are consistent with the remaining three bridging hydrogen atoms taking part in a fluxional process, thereby rendering the four B(basal)–B(basal) bonds equivalent.

The structure of pentaborane(11) is more open than that of pentaborane(9). In essence, B_5H_9 has been reduced in going to B_5H_{11}. This underlines another generalization which we shall explore further in Chapter 6: *reduction of a borane or metalloborane cluster leads to cage opening*. Experimentally, the transformation of B_5H_9 to B_5H_{11} takes place in two steps [10]:

$$B_5H_9 + 2M[C_{10}H_8] \rightarrow M_2[B_5H_9] + C_{10}H_8$$

$$M_2[B_5H_9] + 2HX \rightarrow B_5H_{11} + 2MX$$

$$M = K, Cs; \qquad C_{10}H_8 = naphthalene; \qquad X = Cl, Br$$

The anion, $[B_5H_9]^{2-}$, has been characterized spectroscopically, and the cage is highly fluxional. Even at low temperature, the ^{11}B NMR spectrum exhibits a highfield doublet (one apical boron atom, $\delta^{11}B$ −51.7) and a second doublet (four boron atoms, $\delta^{11}B$ −16.1), despite the fact that the static structure (Fig. 4.5.2) supports three boron atom environments. Protonation of the $[B_5H_9]^{2-}$ ion provides B_5H_{11} in good yield and this route represents the most recent, and most convenient, method of synthesizing B_5H_{11} [10].

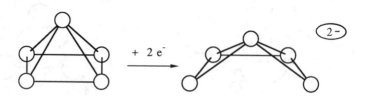

Fig. 4.5.2 — Rearrangement of the B_5-cage which accompanies the reduction of B_5H_9 to $[B_5H_9]^{2-}$.

The structure of pentaborane(11) (Fig. 4.5.3) has been extensively studied by X-ray and electron diffraction methods as well as by microwave spectroscopy. A comparison of diffraction data is given in Table 4.5.2. The open B_5-cage of the B_5H_{11} molecule consists of three, edge-fused B_3-triangles. Each of the two boron atoms of lowest cage connectivity bears two terminal hydrogen atoms; this situation is similar to that found in B_4H_{10}. The most intriguing feature of the structure, however, is the presence of an apparently face-capping hydrogen atom. The precise location of this atom has been determined as a result of an electron diffraction study [8]. The unique hydrogen atom is essentially terminal with respect to the apical boron atom, but leans down towards the open face of the B_5-cage such that an interaction with the two extreme basal boron atoms is allowed. The electron diffraction data imply that this face-capping mode is asymmetrical (Table 4.5.2).

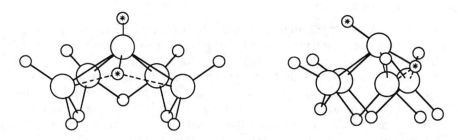

Fig. 4.5.3 — Structure of B_5H_{11} confirmed by X-ray and electron diffraction studies; the second view emphasizes the unusual position of the terminal hydrogen atom associated with the apical boron atom.

Table 4.5.2 — Structural parameters for B_5H_{11} obtained from X-ray [11] and electron diffraction [8] studies

	X-ray diffraction	Electron diffraction
Bond length (Å)		
B(apical)–B(basal)	1.72; 1.87	1.742; 1.892
B(basal)–B(basal)	1.77; 1.75	1.760; 1.812
B(basal)–H(terminal) (av.)	1.15	1.192
B(apical)–H(terminal)	1.02	1.192
B(apical)–H(unique-bridging)	1.09	1.32
B(basal)–H(unique-bridging)	1.67	1.594; 1.899
B(basal)–H(bridge)	1.32	1.394

The $[B_5H_{12}]^-$ anion is prepared by the addition of a BH_3 fragment (added as B_2H_6) to $[B_4H_9]^-$. The structure illustrated in Fig. 4.5.4 has been proposed on the basis of spectroscopic data [6].

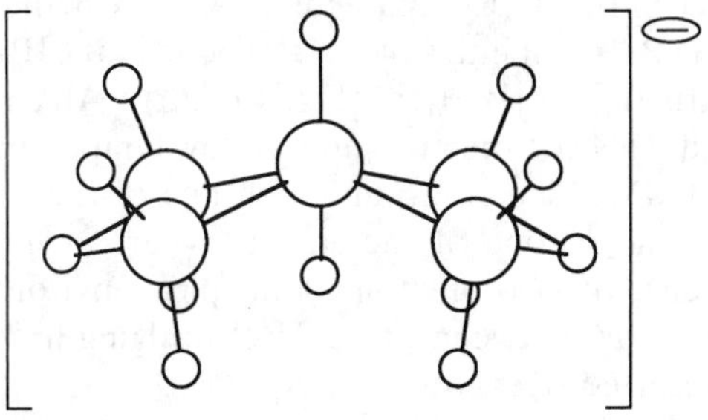

Fig. 4.5.4 — Proposed structure of $[B_5H_{12}]^-$.

4.6 SIX BORON ATOMS

The solid-state structure [12] of hexaborane(10), B_6H_{10}, is depicted in Fig. 4.6.1(a). The B_6-cage is a pentagonal pyramid, with bridging hydrogen atoms along four of the five B(basal)–B(basal) edges. Not surprisingly, in solution, a fluxional process is observed which interchanges all the bridging hydrogen atoms and renders all of the B(basal)–B(basal) bonds equivalent. X-ray diffraction data show that the unique B(basal)–B(basal) bond in B_6H_{10} is remarkably short (1.596 Å), while other B–B distances are as expected with B(apical)–B(basal) distances ranging from 1.736 to 1.795 Å, and hydrogen bridged B(basal)–B(basal) lengths being 1.737 and 1.794 Å.

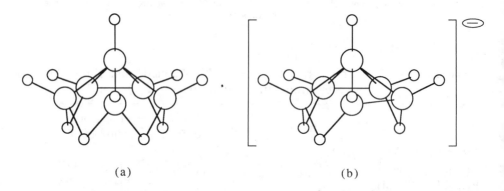

(a)   (b)

Fig. 4.6.1 — (a) Crystallographically determined structure of B_6H_{10}, and (b) proposed structure of the $[B_6H_9]^-$ anion.

As with B_5H_9, deprotonation of B_6H_{10} occurs via the removal of a proton from a B–H–B bridging site. Structural assignment of the $[B_6H_9]^-$ anion has been made on the basis of spectroscopic data and the proposed molecular geometry is represented in Fig. 4.6.1(b) [9]. In solution, the bridging hydrogen atoms exhibit fluxional behaviour, once again rendering the five basal B–B bonds equivalent. For each of B_6H_{10} and $[B_6H_9]^-$, the ^{11}B NMR spectrum shows a characteristically high field resonance for the apical boron atom.

The anion, $[B_6H_{11}]^-$, may be prepared from $[B_5H_8]^-$ by reaction with diborane(6) (i.e. a source of BH_3). (Compare this strategy with the synthesis of $[B_5H_{12}]^-$.) The insertion of the BH_3 unit into a basal B–B bond of the $[B_5H_8]^-$ anion is apparent in the proposed structure of $[B_6H_{11}]^-$ (Fig. 4.6.2(a)). An open cage is suggested on the basis of ^{11}B and ^1H NMR spectroscopy. At low temperature, four resonances are evident in the ^{11}B NMR spectrum, although two of the signals overlap with one another [6]. Once again, one of the signals occurs at high field, and therefore indicates the presence of a boron atom in an apical environment. At temperatures ≥25°C, exchange occurs between the B–H–B bridging hydrogen atoms and those belonging to the inserted BH_3 group.

Protonation of $[B_6H_{11}]^-$ provides an efficient route to hexaborane(12). Twelve years elapsed between the synthesis of this compound [6] and its structural characterization by electron diffraction [13]. Characterization by X-ray crystallography has

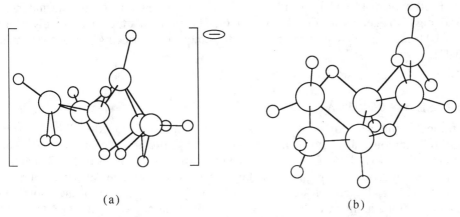

Fig. 4.6.2 — (a) Proposed structure of the $[B_6H_{11}]^-$ anion; (b) structure of B_6H_{12} confirmed by electron diffraction studies.

been prevented because B_6H_{12} forms a glass at low temperature. The structure of B_6H_{12} is closely related to that of B_4H_{10}, and may be viewed by replacing each of two B–H–B bridges by a BH_2-unit; compare Fig. 4.6.2(b) with Fig. 4.4.1. The remaining two B–H–B bridges become particularly asymmetrical, with B–μH distances of 1.20 and 1.42 Å. The central B_4-butterfly in B_6H_{12} is significantly more open than in B_4H_{10}, the internal dihedral angles being 167.4° and 117.4° respectively. Hexaborane(14), B_6H_{14}, has been reported [14] but its structure has yet to be established.

The hexanuclear boron hydrides herald the start of a series of dianions of formula $[B_nH_n]^{2-}$ ($n = 6, 7, 8, 9, 10, 11, 12$), each of which possesses a closed polyhedral cluster of boron atoms.[†] The structure of the first member of this series, the hexahydrohexaborate dianion, $[B_6H_6]^{2-}$, is illustrated in Fig. 4.6.3(a). X-ray diffraction data [15] confirm a regular octahedral array of boron atoms (B–B = 1.69 Å), with each terminal hydrogen atom directed radially with respect to the centre of the octahedron.

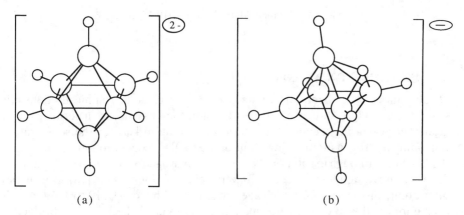

Fig. 4.6.3 — Crystallographically determined structures of (a) $[B_6H_6]^{2-}$ and (b) $[B_6H_7]^-$.

[†] The structures of $[B_7H_7]^{2-}$ and $[B_{11}H_{11}]^{2-}$ have not been confirmed.

Protonation of $[B_6H_6]^{2-}$ yields $[B_6H_7]^-$ which is noteworthy for its unusual face-bridging hydrogen atom (Fig. 4.6.3(b)) [16]. The mode of attachment is truly triply (BBB) bridging, unlike that described for the unique face-bridging hydrogen atom in B_5H_{11}.

4.7   SEVEN BORON ATOMS

Strangely, the class of heptaboranes is represented only by the $[B_7H_7]^{2-}$ and $[B_7H_{12}]^-$ anions, and the former has not been structurally characterized. The structure of $[B_7H_{12}]^-$ (Fig. 4.7.1) has been proposed on the basis of NMR spectral data and by considering the crystallographically determined structure of $[\{\mu\text{-}Fe(CO)_4\}B_7H_{12}]^-$. The cluster geometry reflects the preparation of the heptaborane(12) anion via the insertion of a BH_3 unit into a basal B–B bond of $[B_6H_9]^-$. In solution, the B–H–B bridging hydrogen atoms exchange with those associated with the basally bridging BH_3 unit.

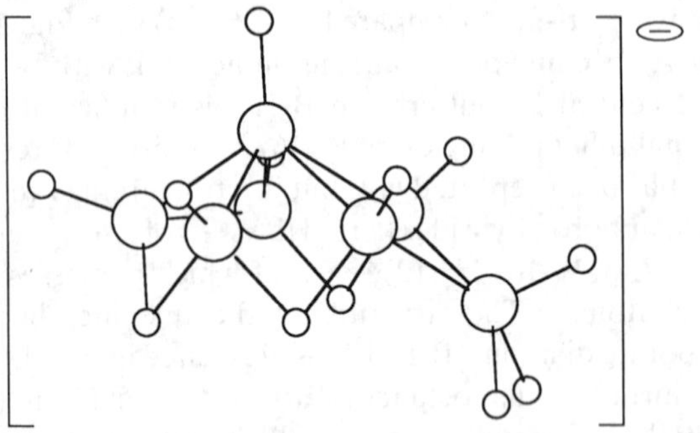

Fig. 4.7.1 — Proposed structure of the $[B_7H_{12}]^-$ anion.

4.8   EIGHT BORON ATOMS

Octaborane(12) has been known since 1964. The crystallographically determined structure of B_8H_{12} is shown in Fig. 4.8.1(a), and consists of an open network of boron atoms, each bearing a single terminal hydrogen atom [17]. In the solid state, the remaining hydrogen atoms bridge four of the six B–B edges around the open face of the B_8-cage, although in solution, two of these hydrogen atoms (those labelled in Fig. 4.8.1(a)) move rapidly between symmetry-related bridging sites [18]. Boron–boron interatomic distances *within* the cage range in length from 1.710 to 1.830 Å, whilst each unbridged B–B edge on the open face of the cage is 1.822 Å. The bridged B–B distances range from 1.674 to 1.806 Å. The X-ray diffraction data available for B_8H_{12} suggest that the hydrogen atoms indicated by the asterisks in Fig. 4.8.1(a) are located asymmetrically in their B–H–B sites.

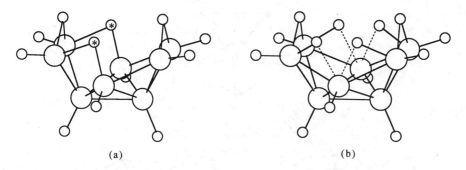

Fig. 4.8.1 — (a) Structure of B_8H_{12} confirmed by X-ray crystallography. The hydrogen atoms involved in the solution fluxional process are labelled *; see text. (b) Proposed structure of B_8H_{14}.

The structure of octaborane(14) has been proposed on the basis of spectroscopic data [19]. Attempts to grow crystals of B_8H_{14} have met with the same problem as we observed earlier for B_6H_{12}, namely that the material forms a glass in the solid state. At $-45°C$, the ^{11}B NMR spectrum of B_8H_{14} in carbon disulphide solution exhibits three signals, with intensity ratios 2:4:2. These data, along with ^1H NMR spectra, are consistent with the structure illustrated in Fig. 4.8.1(b). The B_8-cage is an open one and is clearly related to the one confirmed for B_8H_{12}. Note the occurrence of four 'semi-bridging' hydrogen atoms.

The series of hydroborate dianions continues with $[B_8H_8]^{2-}$. The closed B_8-cage has a dodecahedral geometry (Fig.. 4.8.2(a)) in which four boron atoms have a connectivity equal to 4, and four possess a connectivity number of 5 [20]. (The dodecahedral structure may be viewed as a combination of two B_4-butterflies as illustrated in Fig. 4.8.2(b). The B–B internuclear distances in $[B_8H_8]^{2-}$ vary with atom connectivity, (x): B(4)–B(4) < B(4)–B(5) < B(5)–B(5). In solution, the structure of the $[B_8H_8]^{2-}$ is not so simply represented, since the 8-vertex cage has available to it several polyhedral isomers, all of similar energies. Results of ^{11}B NMR spectroscopic studies indicate that the dodecahedral, the square antiprismatic, and the bicapped trigonal prismatic cages (Fig. 4.8.3) are all accessible to the $[B_8H_8]^{2-}$ anion in solution [21].

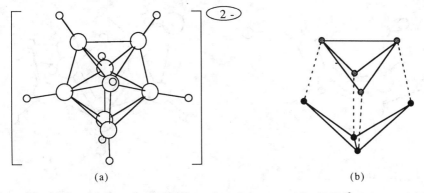

Fig. 4.8.2 — (a) Crystallographically confirmed structure of the $[B_8H_8]^{2-}$ anion, and (b) the generation of a dodecahedral B_8-cluster by fusing of two B_4-butterflies.

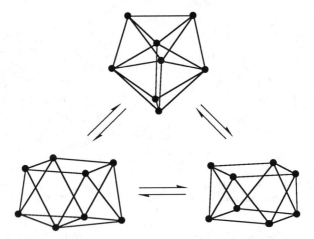

Fig. 4.8.3—Interconversion of the dodecahedron (top), bicapped trigonal prism (bottom left), and square antiprism (bottom right).

4.9 NINE BORON ATOMS

Nonaborane(15) exists in two isomeric forms, of which one, n-B_9H_{15}, has been crystallographically characterized [22]. Although the method of synthesis involves the reaction of B_5H_9 with B_2H_6, the structure of n-B_9H_{15} may be described in terms of the insertion of a BH_3 unit into a B–H–B bridge bond of B_8H_{12}. This may be appreciated by comparing Fig. 4.9.1(a) with Fig. 4.8.1(a); the 'inserted BH_3' group is in the top-right position in Fig. 4.9.1(a). Boron–boron internuclear distances range from 1.75 to 1.95 Å, with the largest separation corresponding to the non-bridged B–B bonds on the open face of the B_9-cage. A second isomer of nonaborane(15), iso-B_9H_{15}, exists. Spectroscopic data suggest the presence of three boron atom environments in a ratio 3:3:3, and the open network illustrated in Fig. 4.9.1(b) has been proposed [19]. This arrangement of boron atoms coincides with that found in the

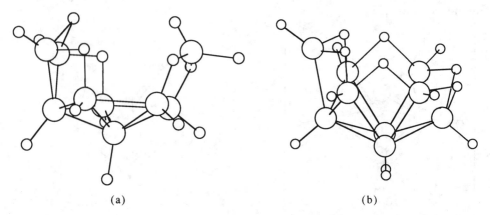

(a)   (b)

Fig. 4.9.1 — (a) Structure of n-B_9H_{15} confirmed by X-ray diffraction studies. (b) Proposed structure of iso-B_9H_{15}.

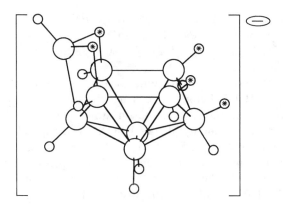

Fig. 4.9.2 — Structure of $[B_9H_{14}]^-$ confirmed by X-ray diffraction studies. Hydrogen atoms marked as * are potential sites for deprotonation.

crystallographically characterized $[B_9H_{14}]^-$ anion (Fig. 4.9.2) [23]. Although the solid-state structure shows the open face of the cage comprising three BH_2 units, three BH units, and two bridging hydrogen atoms, NMR data suggest that in solution, five of these H atoms are fluxional. A proton may be extracted from $[B_9H_{14}]^-$ by using potassium hydride to generate the $[B_9H_{13}]^{2-}$ anion. The B_9-core structure of this dianion is similar to that of $[B_9H_{14}]^-$. The proton removed is one of those indicated by a * in Fig. 4.9.2, and, in the solid state, several isomers of $[B_9H_{13}]^{2-}$ exist. The apparent lack of preference for a site of deprotonation is unusual. In solution, the remaining four H(*) atoms are fluxional [24].

The structure of the $[B_9H_{12}]^-$ anion, confirmed by X-ray diffraction studies, is shown in Fig. 4.9.3. The cage is a fragment of the bicapped Archimedean antiprism (see Fig. 4.1.1.) [25]. Three of the five B–B edges of the open face of the B_9-cage are bridged by hydrogen atoms, although, in solution, these hydrogen atoms are fluxional and are therefore rendered equivalent.

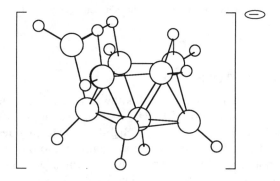

Fig. 4.9.3 — Structures of $[B_9H_{12}]^-$ confirmed by X-ray crystallography.

It is well worth comparing the B_9-cages described above in order to appreciate the close similarities of the core structures of $[B_9H_{12}]^-$, $[B_9H_{14}]^-$ and iso-B_9H_{15}. The formal reduction of the $[B_9H_{12}]^-$ to $[B_9H_{14}]^-$ or $[B_9H_{13}]^{2-}$ anions results in an opening of the B_9-cage along one edge (compare Figs 4.9.2 and 4.9.3). This feature reiterates the point we made concerning the cage opening which accompanies the reduction of B_5H_9 (see Fig. 4.5.2). We shall return to patterns in borane cage structure in Chapter 6.

The nonahydrononaborate dianion, $[B_9H_9]^{2-}$, is a member of the series to which $[B_6H_6]^{2-}$ and $[B_8H_8]^{2-}$, belong. The structure comprises a closed cage, tricapped trigonal prismatic arrangement of boron atoms (Fig. 4.9.4), and there are two different boron environments, dictated by the geometrical requirement for cage connectivity numbers of 4 and 5 [26]. The B(4) atoms are located in the capping sites, whilst the B(5) atoms define the central trigonal prism. The average B(4)–B(5) distance is 1.71 Å whilst the average B(5)–B(5) internuclear separtion is 1.88 Å, and thus as with $[B_8H_8]^{2-}$, we observe that distances between atoms of high connectivity are greater than those involving atoms of low connectivity.

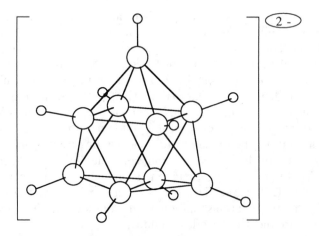

Fig. 4.9.4 — Structure of the $[B_9H_9]^{2-}$ dianion confirmed by X-ray crystallography.

4.10 TEN BORON ATOMS

Decaborane(14) is one of the most well known and commercially important of the boron hydrides. The structure of $B_{10}H_{14}$ is shown in Fig. 4.10.1(a), and has been studied both by X-ray [27] and neutron [28] diffraction techniques. The B_{10}-cage has an open, 6-atom face which exhibits a 'boat' conformation. Four of the six edges of the face are asymmetrically bridged by hydrogen atoms (B–μH = 1.298 and 1.347 Å) [28]. The unbridged B–B edges are significantly longer than the bridged B–B separations as the bond lengths listed in Table 4.10.1 indicate. Conversion of $B_{10}H_{14}$ to $[B_{10}H_{13}]^-$ occurs via the removal of one of the B–H–B bridging protons, and the B_{10}-cage of the anion (Fig. 4.10.1(b)) varies little from that of its conjugate acid (see Table 4.10.1) [29]. On the other hand, the reduction of $B_{10}H_{14}$ to $[B_{10}H_{14}]^{2-}$ results in a rearrangement of the hydrogen atoms (Fig. 4.10.1(c)).

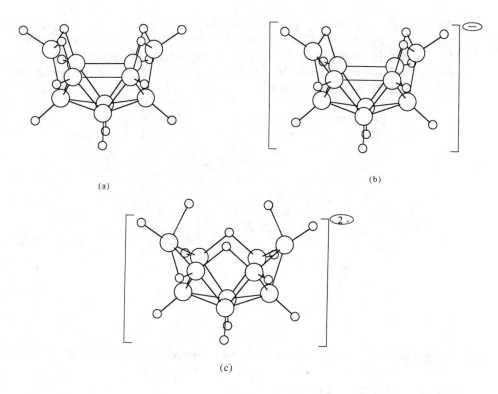

Fig. 4.10.1 — Crystallographically confirmed structures of (a) $B_{10}H_{14}$, (b) $[B_{10}H_{13}]^-$ and (c) $[B_{10}H_{14}]^{2-}$.

Table 4.10.1 — Bond length data for the open face of $B_{10}H_{14}$, $[B_{10}H_{13}]^-$, and $[B_{10}H_{14}]^{2-}$ [28–30]

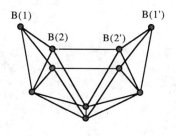

Bond type	$B_{10}H_{14}$	$[B_{10}H_{13}]^-$	$[B_{10}H_{14}]^{2-}$
B(1)–B(2)	1.775	1.65(unbridged) 1.78(bridged)[a]	1.881
B(2)–B(2')	1.973	1.86; 2.04	1.887

[a] Average of 3 edges.

The $[B_{10}H_{10}]^{2-}$ dianion possesses a closed, bicapped, Archimedean antiprismatic framework, (Fig. 4.10.2) [31]. Consistent with the data for both $[B_8H_8]^{2-}$ and $[B_9H_9]^{2-}$, the B–B bond lengths in $[B_{10}H_{10}]^{2-}$ fall into groups according to the atom connectivities. Thus, the average B(4)–B(5) distance in $[B_{10}H_{10}]^{2-}$ is 1.73 Å and the average B(5)–B(5) separation is 1.84 Å.

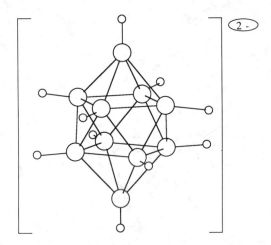

Fig. 4.10.2 — Structure of the $[B_{10}H_{10}]^{2-}$ dianion confirmed by X-ray crystallography.

4.11 ELEVEN BORON ATOMS

Eleven-boron-atom boranes have proved to be extremely elusive, and full structural characterization of such clusters remains limited to the $[B_{11}H_{13}]^{2-}$ and $[B_{11}H_{14}]^{-}$ anions. The framework of the $[B_{11}H_{13}]^{2-}$ dianion is derived from an icosahedron. The open face of the cage (Fig. 4.11.1(a)) is planar, and two of the five edges are bridged by hydrogen atoms; B–B(bridged) = 1.89 Å, and B–B(unbridged) = 1.80 and 1.82 Å [32]. This contrasts sharply with the situation we encountered for $B_{10}H_{14}$ (see Table 4.10.1). Within the B_{11}-cage itself, B–B bond distances range from 1.72 to 1.83 Å.

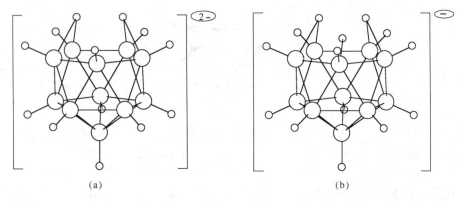

(a)   (b)

Fig. 4.11.1 — Structures of (a) the $[B_{11}H_{13}]^{2-}$ dianion, and (b) the $[B_{11}H_{14}]^{-}$ anion confirmed by X-ray diffraction studies.

The $[B_{11}H_{14}]^-$ anion is the conjuguate acid of $[B_{11}H_{13}]^{2-}$, although it is, in fact, most conveniently prepared from B_5H_9 via treatment with an alkali metal hydride or butyl lithium [33]. The structure of $[B_{11}H_{14}]^-$ is represented in Fig. 4.11.1(b), and bears a close resemblance to that of $[B_{11}H_{13}]^{2-}$. The additional proton, although terminally attached to a boron atom, points in towards the open B_5-face. The B–B distances for the open face range from 1.875 to 1.895 Å, showing little variation between bridged and unbridged bonds. The average distance between adjacent boron atoms within the cage is slightly shorter (1.762 Å) than in $[B_{11}H_{13}]^{2-}$. Protonation of $[B_{11}H_{14}]^-$ yields $B_{11}H_{15}$. Spectroscopic data suggest that the cage structure of this neutral boron hydride is similar to that of its conjugate base. However, the four hydrogen atoms located around the open B_5-face of the cage are fluxional even at low temperatures and no static structure has yet been proposed [33].

The series of closed hydroborate dianion clusters continues with $[B_{11}H_{11}]^{2-}$. This anion has long eluded complete structural characterization. The proposed octadeca-hedral B_{11}-cage (Fig. 4.1.1) is highly fluxional in solution and a static structure has not been frozen out. Evidence to support an octadecahedral arrangement of boron atoms in $[B_{11}H_{11}]^{2-}$ is provided by structural data for derivatives such as $[B_{11}H_{10}(SMe_2)]^-$. This cage is also highly fluxional in solution, showing a single ^{11}B NMR signal for the ten unsubstituted boron atoms in addition to the resonance assigned to the substituted atom [34]. The solid-state structure of the $[B_{11}H_{10}(SMe_2)]^-$ anion is drawn in Fig. 4.11.2. One unusual feature is the apical boron atom which exhibits a connectivity of 6.

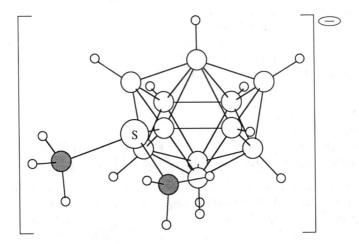

Fig. 4.11.2 — Structure of the $[B_{11}H_{10}(SMe_2)]^-$ anion confirmed by X-ray crystallography.

4.12 TWELVE BORON ATOMS

The regular, icosahedral cage possessed by the $[B_{12}H_{12}]^{2-}$ dianion is shown in Fig. 4.12.1 [35]. The major feature of interest is the high symmetry exhibited by this cage. Each B_5-face has pentagonal symmetry, and the BH units are all equivalent, with each boron–boron interatomic distance being 1.77 Å.

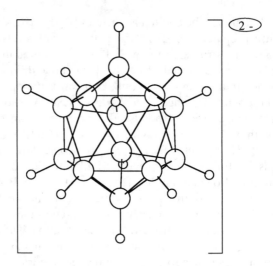

Fig. 4.12.1 — Structure of the $[B_{12}H_{12}]^{2-}$ dianion confirmed by X-ray diffraction studies.

4.13   COUPLED BORANE CAGES

Despite theoretical studies which suggest that closed borane cages in excess of twelve vertices may well be stable, the structures of the experimentally characterized boron hydrides containing thirteen or more boron atoms may conveniently be described as comprising 'coupled units'. Such molecules are called *conjuncto-cages*. Since the scope for coupling borane units together is clearly great, we shall not attempt to cover the area comprehensively.

Coupling of borane cages occurs:

(1) through a vertex with a localized B–B bond, e.g. $\{B_4H_9\}_2$, $\{B_5H_8\}_2$, $\{B_{10}H_{13}\}_2$
(2) through a vertex atom, e.g. $\{B_5H_8\}\{B_2H_5\}$
(3) through a shared B–B edge, e.g. $B_{12}H_{16}$, $B_{13}H_{19}$, $B_{14}H_{18}$, $B_{18}H_{22}$
(3) through a shared BBB face or faces, e.g. $B_{20}H_{16}$.

In addition to the different modes of cage fusion, there also exists the possibility of isomerism as we shall discover in some of the examples below.

For the first group of compounds listed above, cage coupling occurs simply by replacing a terminal hydrogen atom by a direct B–B linkage to a second borane unit. The first example, $B_{10}H_{16}$ (more appropriately formulated as $\{B_5H_8\}_2$), was reported in 1961 [36]. The structure, represented in Fig. 4.13.1, comprises two eclipsed, square pyramidal B_5-cages connected via the apical boron atoms. This isomer is named $1,1'$-$\{B_5H_8\}_2$ (see below, and Fig. 7.1.5 for nomenclature). The B–B bond which joins the two B_5H_8 units together is 1.74 Å in length, a distance which is comparable to the B–B bonds within the cages. In solution, the ^{11}B NMR spectrum of $1,1'$-$\{B_5H_8\}_2$ resembles that of B_5H_9, with, of course, the significant distinction that the high field resonance for the apical boron atom is a singlet in the case of the coupled cluster as opposed to a doublet (^{11}B-^{1}H(terminal) coupling) in

B_5H_9. Two other isomers of $\{B_5H_8\}_2$ have been isolated. Fusion via a basal boron–basal boron linkage provides the 2,2'-$\{B_5H_8\}_2$ isomer, and coupling via a basal boron–apical boron results in the 1,2'-$\{B_5H_8\}_2$ isomer (Fig. 4.13.2). Distinguishing the isomers is again easily achieved by using ^{11}B NMR spectroscopy. For example, the apical boron atoms in 2,2'-$\{B_5H_8\}_2$ are equivalent and are unaffected by cage coupling. Hence, a clean, highfield doublet is observed. In 1,2'-$\{B_5H_8\}_2$, the apical boron atoms are inequivalent, and appear in the ^{11}B NMR spectrum as a singlet and a doublet [37].

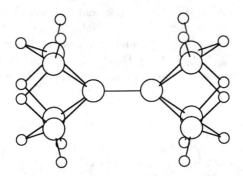

Fig. 4.13.1 — The crystallographically determined structure of 1,1'-$\{B_5H_8\}_2$.

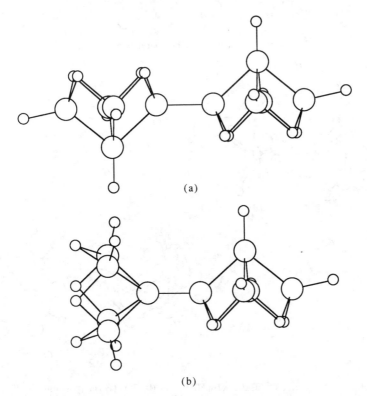

(a)

(b)

Fig. 4.13,2 — Proposed structures of (a) 2,2'-$\{B_5H_8\}_2$ and (b) 1,2'-$\{B_5H_8\}_2$.

The proposed structure of octaborane(18) is shown in Fig. 4.13.3. Spectroscopic data for this borane are similar to those recorded for B_4H_{10} and hence the compound is formulated as $2,2'\text{-}\{B_4H_9\}_2$ [18]. The icosaborane, $B_{20}H_{26}$, is a related example of a coupled cage system. Here, replacement of a terminal hydrogen atom in decaborane(14) by a $B_{10}H_{13}$ unit occurs in several ways to produce isomers of $\{B_{10}H_{13}\}_2$ [38]. The structures of $2,2'\text{-}\{B_{10}H_{13}\}_2$ and $2,6'\text{-}\{B_{10}H_{13}\}_2$ are illustrated in Fig. 4.13.4. $1,5'\text{-}\{B_{10}H_{13}\}_2$ has also been structurally characterized [39]. The individual B_{10}-cages suffer little structural perturbation in going from $B_{10}H_{14}$ to $\{B_{10}H_{13}\}_2$. In each isomer, the B–B bond linking the $B_{10}H_{13}$ units together is rather short (1.692 Å in $2,2'\text{-}\{B_{10}H_{13}\}_2$, 1.679 Å in $2,6'\text{-}\{B_{10}H_{13}\}_2$, and 1.698 Å in $1,5'\text{-}\{B_{10}H_{13}\}_2$), and indicates the presence of a localized 2-centre 2-electron interaction as expected for a terminal bond (see Chapter 6). Although not structurally characterized, the bis-*conjuncto*-borane, $\{B_{10}H_{13}\}\{B_{10}H_{12}\}\{B_{10}H_{13}\}$, is worthy of a brief mention. In theory, 546 possible isomers (including enantiomers) exist! [40].

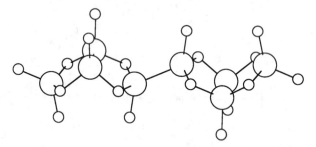

Fig. 4.13.3 — Proposed structure of octaborane(18), or $2,2'\text{-}\{B_4H_9\}_2$.

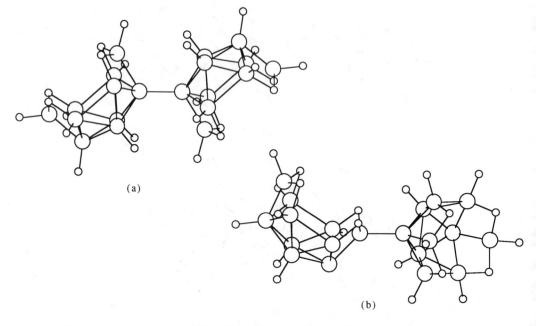

(a)

(b)

Fig. 4.13.4 — Structures of (a) $2,2'\text{-}\{B_{10}H_{13}\}_2$ and (b) $2,6'\text{-}\{B_{10}H_{13}\}_2$ confirmed by X-ray crystallographic studies.

The second type of cage coupling is rather novel, and involves a single boron atom shared between two borane fragments. For example, the transition metal catalysed reaction of B_5H_9 with B_2H_6 gives $2:1',2'-\{B_5H_8\}\{B_2H_5\}$; the nomenclature signifies the involvement of a basal, (2), boron atom in the $\{B_5H_8\}$-cage with both the boron atoms, $(1',2')$, of the $\{B_2H_5\}$-unit. The structure (Fig. 4.13.5) has been proposed on the basis of spectral data [41] and, as with the previous examples we have discussed, a knowledge of the NMR spectroscopic properties of the constituent cages greatly assists in, and adds credence to, the structural assignment.

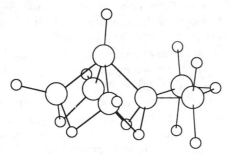

Fig. 4.13.5 — Proposed structure of $2:1',2'-\{B_5H_8\}\{B_2H_5\}$.

Cage coupling via a shared B–B bond is a well exemplified phenomenon. It is a moot point whether we say that a borane cage is a single entity or is composed of two or more coupled units! What constitutes a 'unit'? For instance, consider the structure of B_6H_{12} shown in Fig. 4.6.2(b); could this be described as *conjuncto-*$\{B_4H_6\}\{BH_3\}_2$? The answer here is: 'No, not conveniently'. In fact, a distinction between single and coupled cages arises when we attempt to recognize the deltahedron from which the borane cage is derived. In other words, it may well become convenient to describe any large (usually >12 boron atoms), and often open, cage as a coupled system. The distinction between discrete and coupled cages will become clearer when we discuss the bonding in boron hydrides in Chapter 6.

Dodecaborane(16), $B_{12}H_{16}$, is synthesized by the metal-promoted fusion of two $[B_6H_9]^-$ anions [42]. The structure of the product borane (Fig. 4.13.6) clearly reflects the synthetic origins of the compound, since two pentagonal bipyramidal B_6-cages are apparent. However, it is more convenient to view the structure in terms of an

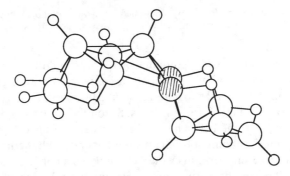

Fig. 4.13.6 — Structure of $B_{12}H_{16}$ confirmed by X-ray diffraction studies; boron atoms not carrying terminal H are shown as shaded circles.

open B_8-cage sharing a B–B edge with a B_6-cage. This conclusion is drawn by noting that *the boron atoms directly involved in cage coupling do not carry terminal hydrogen atoms.* This is a useful criterion to invoke when distinguishing between discrete and coupled cages. In $B_{12}H_{16}$, the shared B–B bond is one of the shortest (1.708 Å) in the entire cage.

The structure of $B_{13}H_{19}$ has been crystallographically characterized and comprises a B_6-cage and a B_9-cage sharing a B–B edge. As in $B_{12}H_{16}$, this linkage is one of the shortest B–B distances (1.703 Å) in the compound. In this case, however, only *one* of the two 'shared' boron atoms meets the criterion of being devoid of a terminal hydrogen atom [43]. Obviously, such atoms are readily distinguished in the ^{11}B NMR spectrum, since no ^{11}B–^{1}H(terminal) spin–spin coupling is observed, and this feature has aided in the structural assignment of $B_{14}H_{18}$. It is proposed that $B_{14}H_{18}$ comprises a decaborane fragment (see Fig. 4.10.1(a)) coupled to a hexaborane cage (see Fig. 4.6.1(a) via a B–B edge [44].

The question of cage isomerism arises for octadecaborane(22). Both n-$B_{18}H_{22}$ and iso-$B_{18}H_{22}$ have been fully characterized, and the structure of the former isomer is illustrated in Fig. 4.13.7. In each isomer, two B_{10}-cages couple via a B–B edge, and iso-$B_{18}H_{22}$ differs from n-$B_{18}H_{22}$ only with respect to the orientation of the two B_{10}-cages [45]. Once again, the boron atoms which are directly involved in cage coupling do not carry terminal hydrogen atoms.

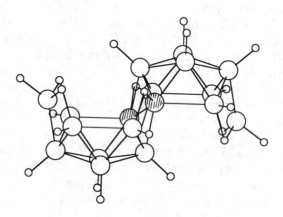

Fig. 4.13.7 — Structure of n-$B_{18}H_{22}$ confirmed by X-ray crystallography; boron atoms not carrying terminal H are shown as shaded circles.

According to our previous guidelines, $B_{14}H_{20}$ should be considered as a 'large cage' and therefore we might anticipate a *conjuncto*-borane. Its structure [46] is illustrated in Fig. 4.13.8 and two features should be noted. Firstly, there are no boron atoms devoid of terminal hydrogen atoms. Secondly, the cage is not as open as those described above. On the other hand, the network of fourteen boron atoms is not readily interpreted as being a fragment of a common deltahedron, and, therefore, the coupling of two smaller units would certainly be a convenient way of describing the structure. The synthetic route to $B_{14}H_{20}$ involves the reaction of $[B_6H_9]^-$ with B_8H_{12} and this fragmentation provides a reasonable description of the final product.

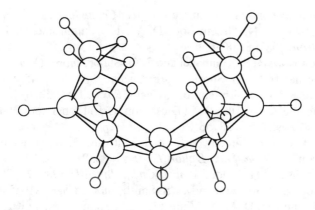

Fig. 4.13.8 — Structure of $B_{14}H_{20}$ confirmed by X-ray diffraction studies.

The coupling of two borane fragments via one of more BBB-faces is an unusual phenomenon. Since, as above, boron atoms residing at the points of cage fusion are denuded of terminal hydrogen atoms, a boron hydride incorporating a face-fused coupled cage has an unusual stoichiometry; there are fewer hydrogen atoms present than boron atoms. An example is $B_{20}H_{16}$, the structure of which is drawn in Fig. 4.13.9. The B_{20}-cluster is a closed cage, and is thus rather unusual among coupled cage systems. It may be viewed as comprising two interpenetrating icosahedra which share a B_4-butterfly unit, indicated by the shaded circles in Fig. 4.13.9 [47]. All hydrogen atoms are terminally bonded.

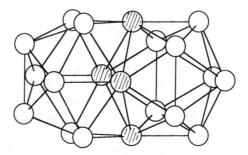

Fig. 4.13.9 — Structure of $B_{20}H_{16}$ confirmed by X-ray diffraction studies. The terminal H atoms, one per B represented by an open circle, have been omitted.

REFERENCES

[1] D. S. Jones and W. N. Lipscomb, *J. Chem. Phys.*, 1969, **51**, 3133.

[2] S. G. Shore and S. H. Lawrence, *J. Am. Chem. Soc.*, 1982, **104**, 7669.

[3] G. F. Mitchell and A. J. Welch, *J. Chem. Soc., Dalton Trans.*, 1987, 1017.

[4] C. E. Nordman and W. N. Lipscomb, *J. Chem. Phys.*, 1953, **51**, 1856.

[5] C. J. Dain and A. J. Downs, *J. Chem. Soc. Dalton Trans.*, 1981, 472.

[6] R. J. Remmel, H. D. Johnson, I. S. Jaworiwsky and S. G. Shore, *J. Am. Chem. Soc.*, 1975, **97**, 5395.

[7] W. J. Dulmage and W. N. Lipscomb, *Acta Crystallogr.*, 1952, **5**, 260.

[8] R. Greatrex, N. N. Greenwood, D. W. H. Rankin and H. E. Robertson, *Polyhedron*, 1987, **6**, 1849.

[9] H. D. Johnson, R. A. Geanagel and S. G. Shore, *Inorg. Chem.*, 1970, **9**, 908.

[10] J. R. Werner and S. G. Shore, *Inorg. Chem.*, 1987, **26**, 1645.

[11] L. Lavine and W. N. Lipscomb, *J. Chem. Phys.*, 1954, **22**, 614.

[12] F. L. Hirshfield, K. Eriks, R. E. Dickerson, E. L. Lippert and W. N. Lispcomb, *J. Chem. Phys.*, 1958, **28**, 56.

[13] R. Greatrex, N. N. Greenwood, M. B. Millikan, D. W. H. Rankin and H. E. Robertson, *J. Chem. Soc., Dalton Trans.*, 1988, 2335.

[14] B. Brellochs and H. Binder, *Angew Chem. Int. Ed.*, 1988, **27**, 262.

[15] R. Schaeffer, Q. Johnson and G. S. Smith, *Inorg. Chem.*, 1965, **4**, 917.

[16] I. Yu. Kuznetsov, D. M. V. Vinitskii, K. A. Solntsev, N. T. Kuznetsov and L. A. Bulman, *Dokl. Akad. Nauk SSSR*, 1985, **283**, 873.

[17] R. E. Enrione, P. F. Boer and W. N. Lipscomb, *J. Am. Chem. Soc.*, 1964, **86**, 1451; *Inorg. Chem.*, 1964, **3**, 1659.

[18] R. R. Reitz, R. Schaeffer and L. G. Sneddon, *Inorg. Chem.*, 1972, **11**, 1242.

[19] D. C. Moody and R. Schaeffer, *Inorg. Chem.*, 1976, **15**, 233.

[20] L. J. Guggenberger, *Inorg. Chem.*, 1969, **8**, 2771.

[21] E. L. Muetterties, *Tetrahedron*, 1974, **30**, 1595.

[22] R. E. Dickerson, P. J., Wheatley, P. A. Howell and W. N. Lipscomb, *J. Chem. Phys.*, 1957, **27**, 200.

[23] N. N. Greenwood, J. A. McGinnety and J. D. Owen, *J. Chem. Soc., Dalton Trans.*, 1972, 986.

[24] T. D. Getman, J. A. Krause, P. M. Niedenzu and S. G. Shore, *Inorg. Chem.*, 1989, **28**, 1507.

[25] G. B. Jacobsen, D. G. Meina, J. H. Morris, S. J. Andrews, D. Reed, A. J. Welch and D. F. Gaines, *J. Chem. Soc., Dalton Trans.*, 1985, 1645.

[26] L. J. Guggenberger, *Inorg. Chem.*, 1968, **7**, 2260.

[27] E. B. Moore, R. E. Dickerson and W. N. Lipscomb, *J. Chem. Phys.*, 1957, **27**, 209.

[28] A. Tippe and W. C. Hamilton, *Inorg. Chem.*, 1969, **8**, 464.

[29] L. G. Sneddon, J. C. Huffman, R. O. Schaeffer and W. R. Streib, *J. Chem. Soc., Chem. Commun.*, 1972, 474.

[30] D. S. Kendall and W. N. Lipscomb, *Inorg. Chem.*, 1973, **12**, 546.

[31] R. D. Dobrott and W. N. Lipscomb, *J. Chem. Phys.*, 1962, **37**, 1779.

[32] C. J. Fritchie, *Inorg. Chem.*, 1967, **6**, 1199.

[33] T. D. Getman, J. A. Krause and S. G. Shore, *Inorg. Chem.*, 1988, **27**, 2398.

[34] E. H. Wong, L. Prasad, E. J. Grabe and M. G. Gatter, *Inorg. Chem.*, 1983, **22**, 1143.

[35] J. A. Wunderlich and W. N. Lipscomb, *J. Am. Chem. Soc.*, 1960, **82**, 4427.

[36] R. Grimes, F. E. Wang, R. Lewin and W. N. Lipscomb, *Proc. Nat. Acad. Sci. USA*, 1961, **47**, 996.

[37] D. F. Gaines, T. V. Iorns and E. N. Clevenger, *Inorg. Chem.*, 1971, **10**, 1096.

[38] S. K. Boocock, N. N. Greenwood, J. D. Kennedy, W. S. McDonald and J. Staves, *J. Chem. Soc., Dalton Trans.*, 1980, 790; S. K. Boocock, Y. M. Cheek, N. N. Greenwood and J. D. Kennedy, *ibid.*, 1981, 1430.

[39] G. M. Brown, J. W. Pinson and L. L. Ingram, *Inorg. Chem.*, 1979, **18**, 1951.
[40] S. K. Boocock, N. N. Greenwood and J. D. Kennedy, *J. Chem. Res.(S)*, 1981, 50.
[41] E. W. Corcoran and L. G. Sneddon, *J. Am. Chem. Soc.*, 1985, **107**, 7446.
[42] C. T. Brewer, R. G. Swisher, E. Sinn and R. N. Grimes, *J. Am. Chem. Soc.*, 1985, **107**, 3558.
[43] J. C. Huffman, D. C. Moody and R. Schaeffer, *Inorg. Chem.*, 1976, **15**, 227.
[44] S. Hermanek, K. Fetter, J. Plesek, L. J. Todd and A. R. Garber, *Inorg. Chem.*, 1975, **14**, 2250.
[45] P. G. Simpson and W. N. Lipscomb, *J. Chem. Phys.*, 1963, **39**, 26; P. G. Simpson, K. Folting, R. D. Dobrott and W. N. Lipscomb, *ibid.*, 1963, **39**, 2339.
[46] J. C. Huffman, D. C. Moody and R. Schaeffer, *Inorg. Chem.*, 1981, **20**, 741.
[47] R. D. Dobrott, L. B. Friedman and W. N. Lipscomb, *J. Chem. Phys.*, 1964, **40**, 866.

5

Structures of metalloboranes

5.1 INTRODUCTION

The number of metalloborane compounds now characterized is vast, and a complete survey is not appropriate for this text. For comprehensive coverage of metalloborane compounds, including structural details, the reader is directed to authoritative accounts by Shore *et al.* [1], Grimes [2], and Kennedy [3]. Two further articles focusing upon metalloboranes with more metal atoms than boron atoms (the so-called *metal-rich metalloboranes*) should also prove valuable [4,5].

In the present chapter, we shall consider the different ways in which boranes interact with main group and transition metal fragments. Metalloborane compounds are, in some cases, best described as metal complexes with the borane functioning as a ligand, e.g. interaction of the borohydride anion, $[BH_4]^-$, with a metal centre. On the other hand, the incorporation of one or more metal fragments into a large borane cage does not usually disrupt the cage, and, therefore, the metalloborane is reasonably described as a cluster. If one is searching for 'rules', then, perhaps, we might say that it is appropriate for a metalloborane containing one, or maybe two, metal atoms and a *small* borane unit (maybe $\leqslant 3$ boron atoms) to be classified as a complex whilst most other systems are more conveniently considered as clusters. This statement is necessarily vague since, as we shall see below, generalizations are not easily made. Certainly, though, when a bonding description is required for a metalloborane in which the boron atom is apparently involved in more than four 'bonds', then the compound is usually treated as a cluster.

5.2 THE BOROHYDRIDE ANION, $[BH_4]^-$, AS A LIGAND

Although the $[BH_4]^-$ anion is isoelectronic with CH_4, its chemistry is significantly different. The borohydride anion is an important reducing agent, being a convenient source of hydrogen:

$$[BH_4]^- \rightarrow BH_3 + \tfrac{1}{2}H_2 + e^-$$

and is regularly used as a source of hydride ion, for instance in the preparation of transition metal hydride compounds:

$$Fe(CO)_4I_2 \xrightarrow{[BH_4]^-} Fe(CO)_4H_2 + 2I^-$$

Coordination of $[BH_4]^-$ to main group, transition, lanthanide, and actinide metals is common [6], although the rôle of $[BH_4]^-$ is a ligand may compete with its activity as a reducing agent. In each of the examples below, reduction of the metal accompanies complex formation:

$$2CpTiCl_3 + 6Li[BH_4] \rightarrow 2CpTi(BH_4)_2 + 6LiCl + B_2H_6 + H_2$$
$$\text{Ti}^{IV} \qquad\qquad\qquad \text{Ti}^{III}$$

$$2Cp_2TiCl_2 + 4Li[BH_4] \rightarrow 2Cp_2Ti(BH_4) + 4LiCl + B_2H_6 + H_2$$
$$\text{Ti}^{IV} \qquad\qquad\qquad \text{Ti}^{III}$$

$$VCl_4(thf)_2 + 5Li[BH_4] \rightarrow Li[V(BH_4)_4] + 4LiCl + \tfrac{1}{2}B_2H_6 + \tfrac{1}{2}H_2 + 2thf$$
$$\text{V}^{IV} \qquad\qquad\qquad \text{V}^{III}$$

On the other hand, since a successful reduction depends upon the relative values of $E^{\ominus}[M^{n+}/M^{(n-1)+}]$ and $E^{\ominus}[BH_3 + \tfrac{1}{2}H_2/[BH_4]^-]$, some complexes will form without an accompanying redox reaction, e.g.

$$HfCl_4 + 4Li[BH_4] \rightarrow Hf[BH_4]_4 + 4LiCl$$

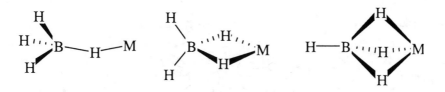

Fig. 5.2.1 — Modes of coordination of the $[BH_4]^-$ ligand to a single metal centre.

The borohydride ligand has three potential modes of coordination to a metal atom (Fig. 5.2.1), conveniently classified as mono-, bi-, or tridentate. In addition, it is capable of bridging between two or three metal centres, as observed in the complexes $(Cp^*Ir)_2H_3(BH_4)^+$ [11] and $[\{MeC\{(CH_2)PPh_2\}_3RuH\}_2(BH_4)]^+$ [12].

$^\dagger Cp^* = C_5Me_5$.

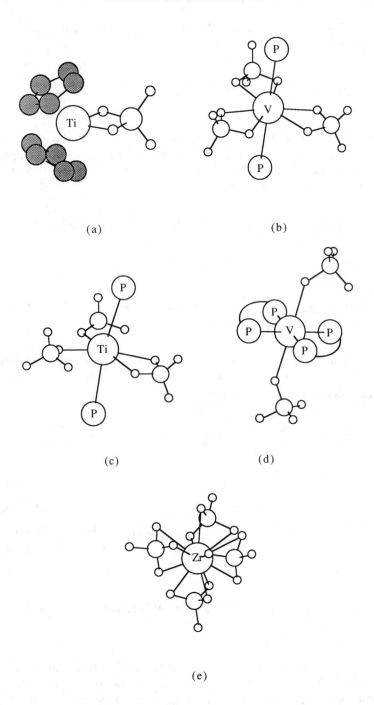

Fig. 5,2.2 — Examples of some complexes involving the borohydride ligand: (a) $Cp_2Ti(BH_4)$ [7], (b) $V(BH_4)_3(PMe_3)_2$ [8], (c) $Ti(BH_4)_3(PMe_3)_2$ [9], (d) $V(BH_4)_2(Me_2PCH_2CH_2PMe_2)_2$ [8], (e) $Zr((BH_4)_4$ [10]. Ⓟ = PMe_3 or PMe_2; H atoms omitted from Cp ring).

The structures of some representative metal complexes are shown in Fig. 5.2.2 (the $Me_2PCH_2CH_2PMe_2$ ligand in Fig. 5.2.2(d) is a chelating phosphine, abbreviated in the diagram to P–P). The modes of coordination of $[BH_4]^-$ may be distinguished by using either diffraction methods or NMR spectroscopy. Even if hydrogen atom location in an X-ray diffraction study fails, the metal–boron internuclear separation should provide some indication of the type of coordination. For instance, in the complex $Ti(BH_4)_3(PMe_3)_2$ (Fig. 5.2.2(c)) the doubly hydrogen bridged Ti–B bond is 2.40 Å in length, whereas each singly bridged bond is 2.27 Å [9]. Contrast this with the situation in the closely related complex, $V(BH_4)_3(PMe_3)_2$ Fig. 5.2.2(b)), in which the V–B bond lengths are all equivalent (2.365 Å). Terminal and bridging hydrogen atoms may, of course, be distinguished in a ^1H NMR spectrum, although for coordinated $[BH_4]^-$ ligands, exchange of the two types of proton is usually facile [6]. For a ligand in which the hydrogen atoms are fluxional in solution at room temperature, but static at low temperatures, the observation in the ^1H NMR spectrum of a single peak (a 1:1:1:1 quartet) which gives rise to two signals of intensity ratio 1:1 at low temperature is indicative of a bidentate mode of coordination.

5.3 DIBORANE- AND TRIBORANE-DERIVED LIGANDS

As soon as we depart from complexes containing the $[BH_4]^-$ ligand, questions arise as to the classification of a metalloborane: is it a complex or a cluster? The compound $CpFe(CO)_2B_2H_5$ possesses a structure which resembles that of diborane, but with an iron fragment replacing one bridging hydrogen atom (Fig. 5.3.1) [13]. An analogous structure is proposed for the related anion, $[Fe(CO)_4B_2H_5]^-$. These complexes may formally be partitioned into the fragments $[CpFe(CO)_2]^+$, or $[Fe(CO)_4]$, and $[B_2H_5]^-$, and since $[B_2H_5]^-$ is isoelectronic* with C_2H_4, the two metalloboranes may be likened to metal alkene complexes [14].

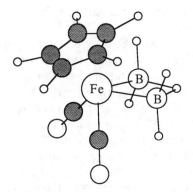

Fig. 5.3.1 — Structure of $CpFe(CO)_2B_2H_5$ confirmed by X-ray crystallography.

---

†Strictly, the $[B_2H_4]^{2-}$ ion is isoelectronic with C_2H_4.

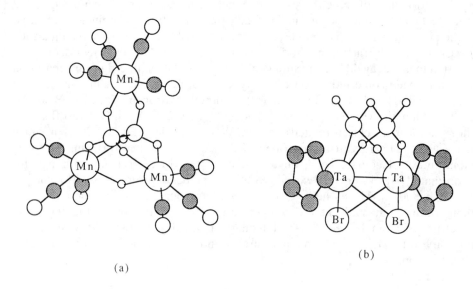

Fig. 5.3.2 — Molecular structures of (a) $HMn_3(CO)_{10}B_2H_6$ and (b) $Cp^*_2Ta_2(\mu\text{-}Br)_2(B_2H_6)$; methyl groups have been omitted from the Cp^* rings.

A diborane ligand is identified in the complexes $Cp^*_2Ta_2(\mu\text{-}Br)_2(B_2H_6)$ [15], and $HMn_3(CO)_{10}B_2H_6$ [16], but, rather than exhibiting the doubly B–H–B bridged structure of the B_2H_6 molecule itself (see Fig. 1.2.1), the ligand turns out to be quite versatile. In $HMn_3(CO)_{10}B_2H_6$ (Fig. 5.3.2(a)), the B_2H_6 ligand exhibits an ethane-like structure, with the borane ligand being hexadentate by virtue of the six Mn–H–B interactions. On the other hand, in $Cp^*_2Ta_2(\mu\text{-}Br)_2(B_2H_6)$ (Fig. 5.3.2(b)), the diborane unit is asymmetrically bound, having 3 Ta–H–B bridges, one direct Ta–B bond, and a B–H–B bridge.

The ability of the terminal hydrogen atoms to participate in bridging interactions to metal atoms in not unique to diborane. The octahydrotriborate$(1-)$ anion is an extremely versatile ligand, with the terminal B–H units functioning rather like the tentacles of an octopus! The structures of $Cr(CO)_4B_3H_8$, $Mn(CO)_3B_3H_8$, and $(\mu\text{-}Br)Mn_2(CO)_6B_3H_8$ [17], shown in Fig. 5.3.3, illustrate this feature. In Chapter 3, we discussed in detail the NMR spectra and fluxional behaviour of the $[B_3H_8]^-$ anion. What happens when the anion is coordinated to a metal atom? The answer is not a simple one, since thermodynamic accessibility of a fluxional process depends upon the mode of coordination of the $[B_3H_8]^-$ ligand, and is affected by other ligands attached to the metal atom as well as by the metal atom itself. For example, $Mn(CO)_4B_3H_8$ is static in solution at room temperature, whereas the hydrogen atoms in $Mn(CO)_3B_3H_8$ undergo facile exchange. Finally, an interesting feature to note as one progresses from left to right across Fig. 5.3.3 is the emergence of a *cage-like* metalloborane structure. Is a metalloborane cluster developing from a metallo-borane complex? A moot point, maybe.

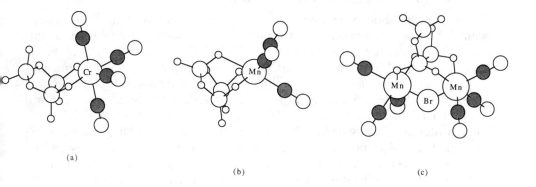

(a)

(b)

(c)

Fig. 5.3.3 — Crystallography determined structures [17] of (a) $Cr(CO)_4B_3H_8$, (b) $Mn(CO)_3B_3H_8$, and (c) $(\mu\text{-Br})Mn_2(CO)_6B_3H_8$.

5.4 METAL FRAGMENTS AS *EXO*-LIGANDS TO BORANE CAGES

In section 4.1, the term *exo* was defined to describe an external hydrogen atom, i.e. a terminal ligand. In rare cases, an *exo*-H atom is replaced by a metal fragment acting as an *exo*-ligand. The oxidative addition of 2-Br–B_5H_8 to *trans*-$IrCl(CO)(PMe_3)_2$ produces $2\text{-}\{IrBr_2(CO)(PMe_3)_2\}(B_5H_8)$, the structure of which is shown in Fig. 5.4.1. (Note that this reaction is accompanied by halide metathesis, hence the change from Ir–Cl to Ir–Br!) The Ir–B bond is relatively short (2.07 Å), and this indicates the presence of a localized 2-centre 2-electron interaction as expected for an *exo*-ligand (see Chapter 6). Despite the steric requirements of the pentaborane unit, the iridium atom exhibits a virtually regular octahedral coordination sphere [18].

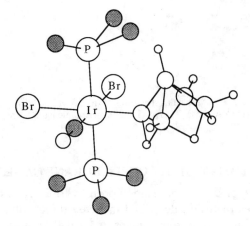

Fig. 5.4.1 — Structure of $2\text{-}\{IrBr_2(CO)(PMe_3)_2\}(B_5H_8)$ confirmed by X-ray diffraction studies; methyl H atoms have been omitted.

5.5 METAL FRAGMENTS AS *ENDO*-LIGANDS TO BORANE CAGES

The term *endo* is used to describe an atom or fragment which is intimately associated with the cluster framework (see Chapter 6), and bridging groups fall into this category. For example, the compound $CpFe(CO)_2B_2H_5$, illustrated in Fig. 5.3.1, possesses a metal fragment, $\{CpFe(CO)_2\}$, which can be considered as an *endo*-group replacing one *endo*-hydrogen atom in B_2H_6. Relationships between *endo*-H atoms and *endo*-metal fragments are often recognized, and two examples are shown in Fig. 5.5.1. The structure of $2,3$-μ-$Cu(PPh_3)_2B_5H_8$ is derived from that of B_5H_9 by the substitution of a basally bridging hydrogen atom by a $Cu(PPh_3)_2$ fragment [19] (the numbering scheme for the B_5-cage is given in Fig. 7.1.5). The copper atom, like each *endo*-hydrogen atom, lies well below the basal-plane of the B_5-square pyramid; this structural feature is rationalized in section 6.4. In $5,6$-μ-$AuP(cyclo$-$C_6H_{11})_3B_{10}H_{13}$, the bonding mode of the $AuP(cyclo$-$C_6H_{11})_3$ fragment mimics that of an *endo*-hydrogen atom in $B_{10}H_{14}$ (for numbering scheme for $B_{10}H_{14}$, see Fig. 7.2.1) [20].

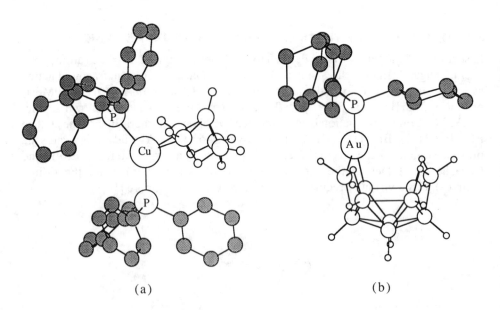

(a)   (b)

Fig. 5.5.1 — Molecular structures of (a) $2,3$-μ-$Cu(PPh_3)_2B_5H_8$ and (b) $5,6$-μ-$AuP(cyclo$-$C_6H_{11})_3B_{10}H_{13}$; H atoms on organic ligands are omitted for clarity.

5.6 METAL ATOMS WHICH COUPLE BORANE CAGES

In Chapter 4, we observed that some boron hydrides consisted of coupled cluster units. The modes of fusion described involved direct intercage B–B bonding, or cages which shared boron atoms, edges or faces. Metal atoms, particularly heavy metals such as silver, gold, mercury, palladium and platinum, can function as centres through which two borane cages couple. In the $[(B_{10}H_{12})_2Au]^-$ anion, illustrated in

Fig. 5.6.1, each $[B_{10}H_{12}]^{2-}$ ligand provides two B–B bonds which function as donor sites and thus, the gold atom (formally Au^{III}) is in a pseudo-square planar environment [21]. A similar geometry is provided for the platinum atom in *trans*-Pt(η^2-$B_6H_{10})_2Cl_2$ (Fig. 5.6.2) [22]. In each case, the metal atom is *endo* with respect to more than one borane cage.

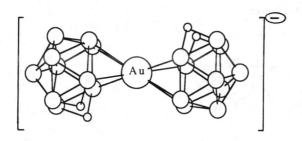

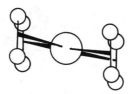

Fig. 5.6.1 — Molecular structure of the $[(B_{10}H_{12})_2Au]^-$ anion; terminal H atoms have been omitted; the scheme at the bottom of the figure emphasizes the psuedo-square planar environment of the Au atom.

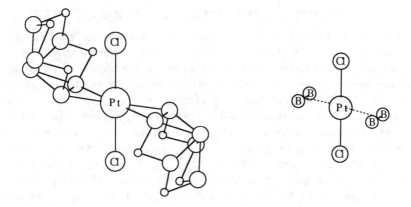

Fig. 5.6.2 — Molecular structure of *trans*-Pt(η^2-$B_6H_{10})_2Cl_2$; terminal H atoms have been omitted. The scheme on the right-hand side emphasizes the pseudo-square planar coordination geometry about the platinum atom.

Coupling of borane cages may involve more than one metal atom, and the metal atoms may or may not be bonded together. In $[B_{10}H_{12}Cd(OEt_2)_2]_2$, each Cd atom is in a pseudo-tetrahedral environment. Each borane fragment is derived from $B_{10}H_{14}$ via the replacement of two *endo*-hydrogen atoms by $Cd(OEt_2)_2$ fragments [23]; (compare Figs 5.6.3(a) and 5.5.1(b)). On the other hand, in $[B_6H_9Pt(PPhMe_2)]_2$, illustrated in Fig. 5.6.3(b), the two metal atoms are within bonding distance of each other (2.644 Å) [24]. Again, each metal atom interacts with the borane cage via a bridging B–M–B mode.

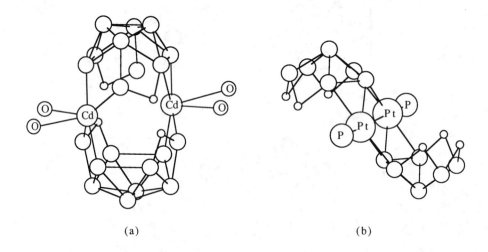

(a)   (b)

Fig. 5.6.3 — Molecular structures of (a) $[B_{10}H_{12}Cd(OEt_2)_2]_2$ and (b) $[B_6H_9Pt(PPhMe_2)]_2$; terminal H atoms and organic substituents have been omitted.

5.7   METAL FRAGMENTS AS CLUSTER UNITS IN BORANE CAGES

In most cases, the interaction of a metal fragment with a borane cage results in the incorporation of the metal atom into the cluster itself. Either the cage may expand, or else formal substitution of metal atom for boron atom occurs. Metal substitution into multinuclear borane cages is potentially complicated by isomerism, although in practice one structure only may be preferred. For example, the structure of $Fe_2(CO)_6B_3H_7$ is derived from pentaborane(9) by replacing each of two BH units by an $Fe(CO)_3$ fragment. Since the square-based pyramidal array of boron atoms in B_5H_9 possesses two unique sites, apical and basal, replacement of *two* atoms could, in theory, lead to *three* isomers of $Fe_2(CO)_6B_3H_7$. These are represented in Fig. 5.7.1(a). Crystallographic characterization in the solid state [25], and spectroscopic characterization in solution [25,26] of $Fe_2(CO)_6B_3H_7$ show that only one isomer is present (Fig. 5.7.1(b)). *Each iron tricarbonyl fragment is an integral part of the cluster.* The fact that, for example, an $Fe(CO)_3$ unit may replace a BH unit in a borane cage without causing massive structural perturbation is rationalized in terms of molecular orbital theory in Chapter 6.

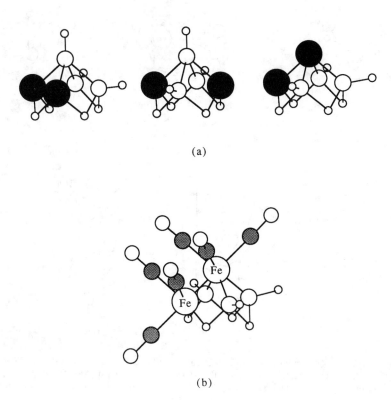

(a)

(b)

Fig. 5.7.1 — (a) The evolution from B_5H_9 of three possible isomers of $Fe_2(CO)_6B_3H_7$ ($\bullet = Fe(CO)_3$); (b) molecular structure of $Fe_2(CO)_6B_3H_7$.

With an ever increasing number of metalloborane clusters appearing in the literature, it is convenient to characterize them as either boron-rich or metal-rich compounds. The structures of the boron-rich metalloboranes are generally related to those of the boron hydrides themselves, while the metal-rich metalloboranes are more readily equated with transition metal clusters. A representative group of structurally characterized boron-rich metalloboranes [27–31] is illustrated in Fig. 5.7.2. The reader should be able to locate a parent borane by comparison with the structures described in Chapter 4. Boron-rich metalloboranes range from low to high cluster nuclearity, and incorporate a variety of coupled cage systems. Often, metal fragments interact with the open face of the parent boron hydride cage, but generalizations in terms of predicting structure should be treated with caution. A comparison of the NMR spectroscopic properties of a metalloborane with those of the parent borane provides valuable information regarding the site of the metal fragment; the ^{11}B NMR resonance for the boron atom which has been replaced by a metal fragment will obviously be absent in the spectrum of the metalloborane derivative.

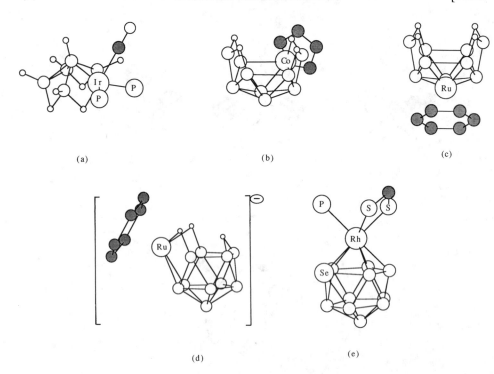

Fig. 5.7.2 — Molecular structures of (a) Ir(PPh$_3$)$_2$(CO)B$_5$H$_8$ (P = PPh$_3$ [27], (b) 5-(η^5-Cp)Co-B$_9$H$_{13}$ [28], (c) 1-(η^6-C$_6$Me$_6$)Ru-B$_9$H$_{13}$ [29], (d) [7-(η^6-C$_6$Me$_6$)Ru-B$_{10}$H$_{13}$]$^-$ [30], (e) 2-{Rh(PPh$_3$)(η^2-S$_2$CH)-1-Se-B$_{10}$H$_{10}$} (P = PPh$_3$) [31]; terminal H atoms have been omitted in structures (b)–(e).

An interesting pair of related compounds is {(η^5-Cp)Co}$_4$B$_4$H$_4$ and {(η^5-Cp)Ni}$_4$B$_4$H$_4$; each exhibits a closed-cage, dodecahedral structure. Remember that the dodecahedron possesses four vertices of connectivity 4, and four of connectivity 5 (Fig. 4.1.1). In {(η^5-Cp)Co}$_4$B$_4$H$_4$, the metal fragments occupy the vertices of *high* connectivity, whilst in {(η^5-Cp)Ni}$_4$B$_4$H$_4$, the nickel atoms reside at the positions of *low* connectivity. Since, in a polyhedral framework, low connectivity corresponds to high electron density (and vice versa), the difference in structure between {(η^5-Cp)Co}$_4$B$_4$H$_4$ and {(η^5-Cp)Ni}$_4$B$_4$H$_4$ may be rationalized in terms of the nickel atom providing a greater number of valence electrons than the cobalt [32].

In section 4.5, we noted that cage opening accompanies cluster reduction, for example in going from B$_5$H$_9$ to [B$_5$H$_9$]$^{2-}$ or B$_5$H$_{11}$. The same phenomenon is noted for metalloborane clusters and is beautifully represented in the series of iridaboranes, IrH(PMe$_3$)$_2$(B$_8$H$_7$Cl), Ir(CO)(PMe$_3$)$_2$(B$_8$H$_{11}$), and Ir(CO)H(PMe$_3$)$_2$(B$_8$H$_{11}$Cl). In crossing this series, each step is accompanied by the *formal* addition of electrons [33]; the *exo*-chlorine ligand is equivalent to an *exo*-hydrogen atom in terms of available valence electrons. In order to emphasize formal reduction is crossing the series, consider the compounds to be formally analogous to the set of clusters HIr(PMe$_3$)$_2$(B$_8$H$_8$), HIr(PMe$_3$)$_2$(B$_8$H$_{12}$), and HIr(PMe$_3$)$_2$(B$_8$H$_{14}$) (see Chapter 6). The stepwise opening of the IrB$_8$-cage which accompanies the reduction is illustrated in Fig. 5.7.3.

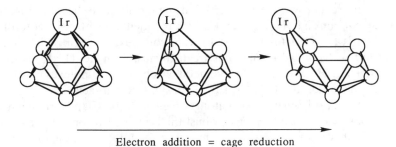

Electron addition = cage reduction

Fig. 5.7.3 — The stepwise of the IrB_8-cage which occurs along the series $IrH(PMe_3)_2(B_8H_7Cl)$, $Ir(CO)(PMe_3)_2(B_8H_{11})$ to $Ir(CO)H(PMe_3)_2(B_8H_{11}Cl)$.

Metal-rich metalloboranes are characterized by possessing a greater number of metal than boron atoms. In the extreme, the borane fragment is stripped down until it becomes a single, naked boron atom surrounded by a metal framework, and at this stage the compound is termed a *metal boride cluster*. Typically, a metal-rich metalloborane retains at least a trinuclear metal fragment, as the examples shown in Fig. 5.7.4 illustrate.

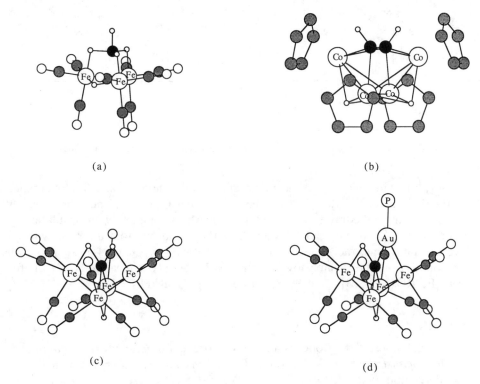

(a)   (b)

(c)   (d)

Fig. 5.7.4 — The molecular structures of some metal-rich metalloboranes: (a) $HFe_3(CO)_9BH_4$ [34], (b) $H_2\{(\eta^5\text{-}Cp)Co\}_4B_2H_2$ [35] (H atoms omitted from the Cp rings), (c) $HFe_4(CO)_{12}BH_2$ [36] and (d) $HFe_4(CO)_{12}\{AuP(cyclo\text{-}C_6H_{11})_3\}BH$ [37], (P = $P(cyclo\text{-}C_6H_{11})_3$).

The astute reader will have noticed that the ferraborane shown in Fig. 5.7.4(a) apparently contains a $[BH_4]^-$ 'ligand' which is tridentate with respect to the metal triangle. As is the case for the borohydride complexes described in section 5.2, terminal and bridging hydrogen atoms are distinguished in the ^1H NMR spectrum of $HFe_3(CO)_9BH_4$. However, *unlike* the metal borohydride complexes which we have previously encountered, $HFe_3(CO)_9BH_4$ does *not* exhibit facile exchange between the terminal and bridging hydrogen atoms. Even at $\approx 100°C$, the terminal hydrogen atom remains *exo* to the Fe_3B-cage, whilst the bridging hydrogen atoms, including the Fe–H–Fe hydride, *are* fluxional. This, along with reactivity characteristics (see section 7.5), implies that the $[BH_4]^-$ moiety is *not* in fact behaving as a borohydride ligand, but is an integral part of the metalloborane cluster. The same is true in the cluster $HRu_3(CO)_9BH_4$ [38].

The characterization of high nuclearity metalloboranes is aided by the use of ^{11}B NMR spectroscopy. As the number of *direct* (i.e. non-bridged) boron–metal interactions increases, the ^{11}B NMR spectral chemical shift for the metal-associated boron atom moves progressively to lower field [4,39]. Examples are shown in Fig. 5.7.5.

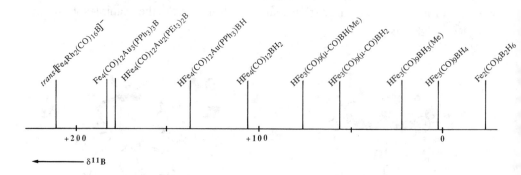

Fig. 5.7.5 — Range of ^{11}B NMR spectral chemical shifts for selected ferraboranes.

Only a few metalloborides have been structurally characterized. Two examples are the $[Fe_4Rh_2(CO)_{16}B]^-$ anion and $Fe_4(CO)_{12}\{Au(PPh_3)\}_3B$. The former cluster consists of an octahedral array of metal atoms which completely encapsulates the boron atom [40]. In $Fe_4(CO)_{12}\{Au(PPh_3)\}_3B$, the boron atom interacts with all seven metal atoms, the metal framework itself being irregular [41]. Again, both compounds exhibit extremely lowfield ^{11}B NMR spectral chemical shifts ($\delta + 211$ for *trans*-$[Fe_4Rh_2(CO)_{16}B]^-$ and $+ 183$ for $Fe_4(CO)_{12}\{Au(PPh_3)\}_3B$).

REFERENCES

[1] K. B. Gilbert, S. K. Boocock and S. G. Shore, in *Comprehensive Organometal-lic Chemistry*, ed. G. Wilkinson, F. G. A. Stone and E. W. Abel, Vol. 6, p. 879, Pergamon, Oxford, 1982.

[2] R. N. Grimes, ed. *Metal Interactions with Boron Clusters*, Plenum Press, New York, 1982.

[3] J. D. Kennedy, *Prog. in Inorg. Chem.*, 1984, **32**, 519; *ibid.*, 1986, **34**, 211.

[4] C. E. Housecroft, *Polyhedron*, 1987, **6**, 1935.

[5] T. P. Fehlner, *New J. Chem.*, 1988, **12**, 307.

[6] T. J. Marks and J. R. Kolb, *Chem. Rev.*, 1977, **77**, 263.

[7] E. Wiberg and H. Neumaier, *Inorg. Nucl. Chem. Lett.*, 1965, **1**, 35.

[8] J. A. Jensen and G. S. Girolami, *J. Am. Chem. Soc.*, 1988, **110**, 4450.

[9] J. A. Jensen, S. R., Wilson and G. S. Girolami, *J. Am. Chem. Soc.*, 1988, **110**, 4977.

[10] V. Plato and K. Hedberg, *Inorg. Chem.*, 1971, **10**, 590.

[11] T. M. Gilbert, F. J. Hollander and R. G. Bergman, *J. Am. Chem. Soc.*, 1985, **107**, 3508.

[12] L. F. Rhodes, L. M. Venanzi, C. Sorato, and A. Albinati, *Inorg. Chem.*, 1986, **25**, 3335.

[13] R. L. DeKock, P. Deshmukh, T. P. Fehlner, C. E. Housecroft, J. S. Plotkin and S. G. Shore, *J. Am. Chem. Soc.*, 1982, **104**, 815.

[14] T. P. Fehlner and C. E. Housecroft, *Adv. Organomet. Chem.*, 1982, **21**, 57.

[15] C. Ting and L. Messerle, *J. Am. Chem. Soc.*, 1989, **111**, 3449.

[16] H. D. Kaesz, W. Fellmann, G. R. Wilkes and L. F. Dahl, *J. Am. Chem Soc.*, 1965, **87**, 2755.

[17] (a) L. J. Guggenberger, *Inorg. Chem.*, 1970, **9**, 367; (b) M. W. Chen, D. F. Gaines and L. G. Hoard, *Inorg. Chem.*, 1980, **19**, 2989; (c) D. F. Gaines, S. J. Hilderbrandt and J. C. Calabrese, *Inorg. Chem.*, 1978, **17**, 790.

[18] M. R. Churchill and J. J. Hackworth, *Inorg. Chem.*, 1975, **14**, 2047.

[19] N. N. Greenwood, J. A. Howard, and W. S. McDonald, *J. Chem. Soc., Dalton Trans.*, 1977, 37.

[20] A. J. Wynd, A. J. McLennan, D. Reed and A. J. Welch, *J. Chem. Soc., Dalton Trans.*, 1987, 2761.

[21] A. J. Wynd and A. J. Welch, *J. Chem. Soc., Chem. Commun.*, 1987. 1174.

[22] J. P. Brennan, R. Schaeffer, A. Davison and S. S. Wreford, *J. Chem. Soc., Chem. Commun.*, 1973, 354.

[23] N. N. Greenwood, J. A. McGinnety and J. D. Owen, *J. Chem. Soc., Dalton Trans.*, 1972, 989.

[24] N. N. Greenwood, M. J. Hails, J. D. Kennedy and W. S. McDonald, *J. Chem. Soc., Chem. Commun.*, 1980, 37.

[25] E. L. Anderson, K. J. Haller and T. P. Fehlner, *J. Am. Chem. Soc.*, 1979, **101**, 4390.

[26] C. E. Housecroft, *Inorg. Chem.*, 1986, **25**, 3108.

[27] N. N. Greenwood, J. D. Kennedy, W. S. McDonald, D. Reed and J. Staves, *J. Chem. Soc., Dalton Trans.*, 1979, 117.

[28] J. R. Pipal and R. N. Grimes, *Inorg. Chem.*, 1977, **16**, 3251.

[29] M. Bown, X. L. R. Fontaine, N. N. Greenwood, J. D. Kennedy and P. MacKinnon, *J. Chem. Soc., Chem. Commun.*, 1987, 817.

[30] M. Bown, X. L. R. Fontaine, N. N. Greenwood, J. D. Kennedy and M. Thornton-Pett, *J. Chem. Soc., Dalton Trans.*, 1987, 1169.

[31] G. Ferguson, Faridoon and T. R. Spalding, *Acta Crystallogr.*, 1988, **44C**, 1368.

[32] M. E. O'Neill and K. Wade, *Inorg. Chem.*, 1982, **21**, 464.

[33] Ref. [3] and references therein.

[34] J. C. Vites, C. E. Housecroft, C. Eigenbrot, M. L. Buhl, G. J. Long and T. P. Fehlner, *J. Am. Chem. Soc.*, 1986, **108**, 3304.

[35] T. P. Fehlner, J. Feilong and A. L. Rheingold, *J. Am. Chem. Soc.*, 1987, **109**, 1860.

[36] T. P. Fehlner, C. E. Housecroft, W. R. Scheidt and K. S. Wong, *Organometallics*, 1983, **2**, 825.

[37] C. E. Housecroft, A. L. Rheingold and M. S. Shongwe, 1989. Unpublished results.

[38] A. K. Chipperfield and C. E. Housecroft, *J. Organometal. Chem.*, 1988, **349**, C17.

[39] N. P. Rath and T. P. Fehlner, *J. Am. Chem. Soc.*, 1988, **110**, 5345.

[40] T. P. Fehlner, R. Khattar, J. Puga and A. L. Rheingold, *J. Am. Chem. Soc.*, 1989, **111**, 1877.

[41] C. E. Housecroft, A. L. Rheingold and M. S. Shongwe, *J. Chem. Soc., Chem. Commun.*, 1988, 965.

6

Bonding in boranes and metalloboranes

6.1 INTRODUCTION

In Chapter 2, we considered the bonding in monoborane and diborane(6), and in Lewis base adducts of BH_3. We introduced the idea that a Lewis base–acid interaction may occur between two molecules of BH_3 (see section 2.3) to generate the diborane molecule. In order to rationalize the structures of the boranes and metalloboranes described in Chapters 4 and 5, we must extend the concept of 'shared electron density'. For example, remembering that the boron atom possesses only three valence electrons, a bonding description for the seven coordinate boron atom in $Fe_4(CO)_{12}\{Au(PPh_3)\}_3B$ (Fig. 6.1.1) is rather difficult to envisage! And yet the compound is stable and so the boron atom must be capable of entering into multi-centre bonding. A molecular orbital treatment is most appropriate for a discussion of the bonding within any cluster compound, but first, let us familiarize ourselves with methods of treating the bonding in simple systems, viz. B_2H_6 and the $[B_3H_8]^-$ ion.

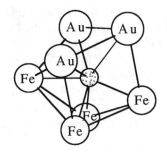

Au = AuPPh3   Fe = Fe(CO)3

Fig. 6.1.1 — The Fe_4Au_3B-core of $Fe_4(CO)_{12}\{Au(PPh_3)\}_3B$ illustrating the 7-coordinate nature of the boron atom.

6.2 DIBORANE(6)

In section 2.3, a molecular orbital description of diborane was given in which we considered the combination of two BH_3 fragments. After distortion to a pyramidal geometry, an acceptor orbital on the boron atom emerged which possessed a suitable symmetry and orientation to interact with a B–H$_{(terminal)}$ bonding orbital on a second BH_3 molecule to generate a B–H–B bridge bonding MO (see Figs 2.3.3–2.3.5). The bonding description did *not* assume a hybridization scheme, but allowed the combination of suitable atomic orbitals of the boron and hydrogen atoms. This is making use of an *atomic orbital basis set*. However, using an atomic orbital basis set is not always convenient, and this is particularly true for large molecules. Although the experimentally determined geometry of the hydrogen atoms around each boron atom in B_2H_6 (see Table 3.5.1) is approximately tetrahedral, some distortion does exist. The H$_{(terminal)}$–B–H$_{(terminal)}$ angle is $\approx 122°$ while the H$_{(bridge)}$–B–H$_{(bridge)}$ angle is $90°$. As far as the terminal hydrogen atoms are concerned, each boron atom should by sp^2 hybridized, although sp^3 hybridization would be more appropriate for the bridging hydrogen atoms. Therefore, it is constructive to consider each hybridization scheme independently and assess the chemical significance of the results.

Let us begin with an sp^3 model. A molecular orbital correlation diagram for B_2H_6 starting from two sp^3 hybridized boron atoms and six hydrogen $1s$ AOs is represented in Fig. 6.2.1, and the construction of the molecular orbitals is illustrated pictorially in Fig. 6.2.2. In the diborane molecule, the approximate tetrahedral environment of each boron atom roughly coincides with the tetrahedral array of four sp^3 hybrid orbitals (Fig. 6.2.2(a)). Per boron atom, there is orbital overlap between two sp^3 hybrids and two hydrogen $1s$ AOs as shown in Fig. 6.2.2(b). Remember that each bonding interaction has an antibonding counterpart, and so, for the B_2H_6 molecule as a whole, four boron sp^3 hybrids interact with four hydrogen $1s$ AOs to generate 4 σ_{BH} and 4 σ^*_{BH} MOs (Ψ_{1-4} and Ψ_{11-14} in Fig. 6.2.1). The remaining four sp^3 hybrid orbitals (two per boron atom) combine to form a B–B bridging interaction, and may overlap in or out-of-phase with one another (Fig. 6.2.2(c)); in considering the different ways in which these sp^3 can combine, we are, in fact, constructing a set of orbitals for two *approaching but nonbonded* BH_2 fragments which have the potential of interacting with two bridging hydrogen atoms to produce B_2H_6. These orbitals are called *ligand group orbitals*, LGOs, and their formation is shown on the left-hand side of Fig. 6.2.1 and in Fig. 6.2.2(c). Since we begin with four sp^3 hybrid orbitals, we must construct four LGOs; the same rules apply to the generation of ligand group orbitals as to any MO formation. Of the four possible LGOs, only two exhibit boron - - - boron bonding character. Note that the interaction is *not* directly along the B–B vector, and could be described as a 'bent' bond between the boron atoms. The remaining two LGOs are antibonding between the two boron atoms.

Now look at the right-hand side of Fig. 6.2.1; of the six hydrogen $1s$ AOs, four have been used for BH(terminal) bonding, and the two remaining $1s$ orbitals will form the B–H–B bridges. By taking these AOs either in or out-of-phase, we generate the LGOs of an H - - - H unit; of course this is similar to forming an H_2 molecule, but in this case, the hydrogen atoms are at a nonbonded separation. In the centre of Fig. 6.2.1, the LGOs for the H_2B - - - BH_2 unit are correlated with those of the H - - - H fragment. (The reason for considering LGOs at all is to restrict the correlations to those between *two* fragments, since diagrams are most easily drawn in 2 dimensions!) The two LGOs of B---B bonding character possess appropriate symmetry to interact with the LGOs of the H---H unit; the result is the formation of two molecular orbitals with B–H–B bonding character, Ψ_5 and Ψ_6, and two with B–H–B antibonding character, Ψ_9 and Ψ_{10}. Interaction of the two LGOs of the H_2B --- BH_2 fragment which possess B---B antibonding character with the H $1s$ orbitals is disallowed by symmetry (look back at Fig. 2.1.7). Hence, Ψ_7 and Ψ_8 of B_2H_6 are nonbonding with respect to the B–H–B interaction.

Once an MO diagram has been constructed, *and not before*, electrons can be added. According to the Aufbau Principle, the lowest energy MOs will be populated. Each boron atom provides 3 valence electrons and each H atom provides one, making a total of 6 electron pairs in all. In Fig. 6.2.1, we observe that this is a sufficient number of electrons to completely occupy the 4 BH(terminal) bonding MOs and the 2 BHB bridge bonding MOs. The sp^3 model therefore implies that there is no direct bonding interaction along the B–B vector, and the diborane molecule appears to be stabilized completely by the B–H–B interactions.

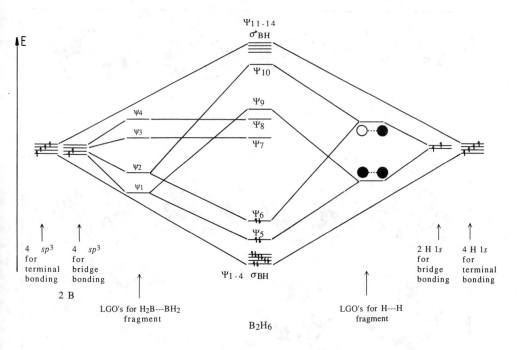

Fig. 6.2.1 — An orbital correlation diagram for the formation of B_2H_6 using an sp^3 model and a ligand group orbital approach; pictorial representations of the molecular orbitals in B_2H_6 are given in Fig. 6.2.2.

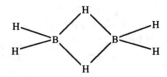

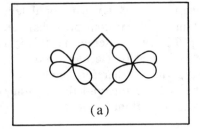

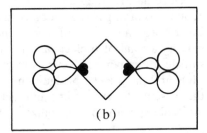

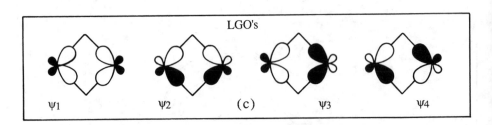

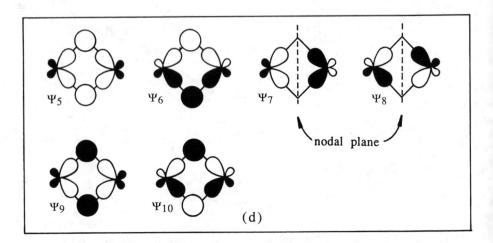

Fig. 6.2.2 — Bonding in diborane described within the confines of an sp^3 model: (a boron sp^3 hybrid orbitals (4 per B atom); (b) formation of terminal B–H bonds; (c) boron sp^3 orbital combinations available for bridge bounding; (d) formation of the B–H–B bridge bonding, nonbonding, and antibonding MOs.

Now, consider an sp^2 model for the bonding in diborane(6). A molecular orbital correlation diagram for B_2H_6 starting from sp^2 hybridized boron atoms and six hydrogen $1s$ AOs is represented in Fig. 6.2.3, and the construction of the molecular orbitals is illustrated pictorially in Fig. 6.2.4. By invoking sp^2 hybridization for each boron atom in B_2H_6 (Fig. 6.2.4(a)), we can satisfactorily account for the $H_{(terminal)}-B-H_{(terminal)}$ angle of $\approx 122°$. The formation of localized $B-H_{(terminal)}$ bonds via the orbital overlap of boron sp^2 hybrids and hydrogen $1s$ AOs is illustrated in Fig. 6.2.4(b); four σ_{BH} and four σ_{BH}^* MOs are formed (Ψ_{1-4} and Ψ_{11-14} in Fig. 6.2.3).

Each boron atom is now left with an sp^2 hybrid orbital which points to the centre of the B–B vector, and an unhybridized $2p$ orbital which lies in the plane of the B–H–B bridges. Using the same stepwise approach as before, we can construct four LGOs for the $H_2B\text{-}\text{-}\text{-}BH_2$ unit by combining orbitals on adjacent boron atoms either in or out-of-phase (Fig. 6.2.4(c) and the left-hand side of Fig. 6.2.3). Immediately, we observe the generation of an MO with *direct* B–B bonding character. This provides a sharp contrast to the sp^3 bonding model. Of the four combinations drawn in Fig. 6.2.4(c), two are of appropriate symmetry to combine with the LGOs of the H - - - H unit and this is illustrated in the centre of Fig. 6.2.3 and in Fig. 6.2.4(d) by the formation of the bonding MOs Ψ_5 and Ψ_6 and their antibonding counterparts, Ψ_9 and Ψ_{10}. The two $H_2B\text{-}\text{-}\text{-}BH_2$ LGOs with B - - - B antibonding character are unable to interact with the bridging H atoms, and remaining nonbonding (Ψ_7 and Ψ_8) with respect to the B–H–B interaction in B_2H_6. The six bonding orbitals in B_2H_6 are then filled with the available electrons.

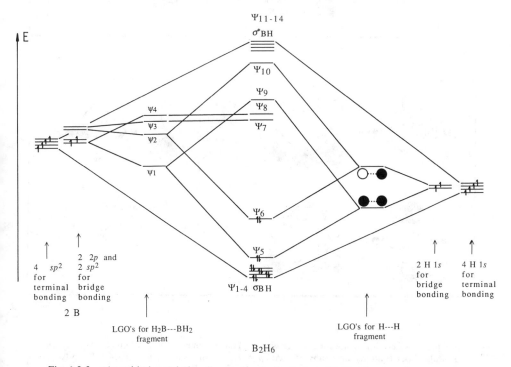

Fig. 6.2.3 — An orbital correlation diagram for the formation of B_2H_6 using an sp^2 model and a ligand group orbital approach; the MOs are represented schematically in Fig. 6.2.4.

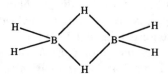

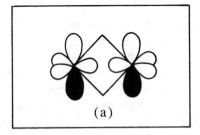

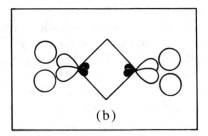

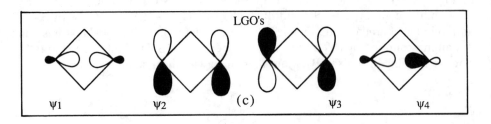

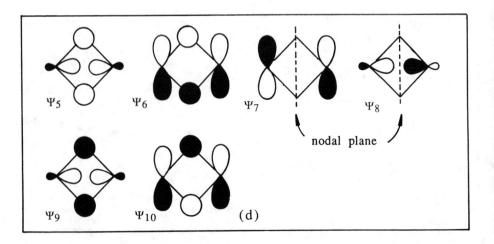

Fig. 6.2.4 — Bonding in diborane described within the confines of an sp^2 scheme: (a) sp^2 hybrid orbitals (3 per B atom), and unhybridized $2p$ AOs; (b) formation of terminal B–H bonds; (c) boron sp^2 and $2p$ orbital combinations available for bridge bonding; (d) formation of the B–H–B bridge bonding, nonbonding, and antibonding MOs.

Which of the two models described above ought we to use? Are we justified in using a hybridization scheme at all? Indeed, is it possible to use only one bonding scheme? Look again at the results presented in section 2.3, and compare Fig. 2.3.4 with Fig. 6.2.1 and 6.2.3. One important observation is that whichever bonding model is used, the total number of B–H–B bonding, nonbonding, and antibonding MOs is the same. Using the atomic orbital basis set, we generated two MOs in B_2H_6 with B–H–B bridge bonding character (Ψ_3 and Ψ_6 in Fig. 2.3.4). Similarly, in both the sp^2 and the sp^3 models, two B–H–B bonding MOs were generated. The molecular orbitals illustrated in Fig. 2.3.4 exhibit characteristics of *both* sp^3 and sp^2 hybridization. The boron atom, like any other, is not restricted to what *we* regard as a conventional hybridization scheme. In reality, the bonding in the B_2H_6 molecule should be considered as a mixture of the sp^3 and sp^2 bonding models: a hybrid of hybrids. On its own, neither model is sufficient, but each helps to rationalize some of the experimentally observed properties of the B_2H_6 molecule.

6.3 THE [B₃H₈]⁻ ANION

The structure of the $[B_3H_8]^-$ anion is reminiscent of that of cyclopropane (Fig. 6.3.1), except, of course, that the hydroborate anion exhibits additional hydrogen atom bridges. In a saturated hydrocarbon molecule, each carbon atom is considered to be sp^3 hybridized. Cyclopropane is a special case because the small ring size results in inefficient overlap of the sp^3 orbitals.

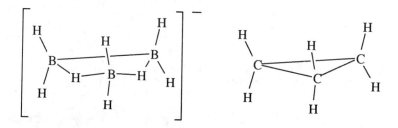

Fig. 6.3.1 — Structures of $[B_3H_8]^-$ and C_3H_6.

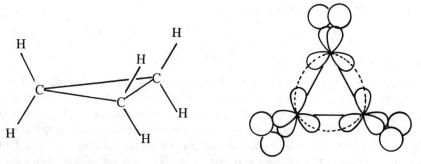

Fig. 6.3.2 — Orbital overlap between carbon and hydrogen atoms in C_3H_6 using sp^3 hybridized carbon atoms. Note that the angle between carbon atoms is 60° while the angle between hybrid orbitals on a given C atom is 109°.

The diagram in Fig. 6.3.2 illustrates that the C–C bonds are 'bent' (dotted lines in Fig. 6.3.2). The hypothetical $[B_3H_6]^{3-}$ trianion is isoelectronic with, and should be isostructural with, C_3H_6. During the protonation of the trianion

$$[B_3H_6]^{3-} + 2H^+ \rightarrow [B_3H_8]^-$$

the protons will be attracted towards regions of available electron density, i.e. the 'bent' B–B bonds; compare this with the 'bent' B–B bonds in the sp^3 model for bonding in B_2H_6 in Fig. 6.2.2(c). The protonation of $[B_3H_6]^{3-}$ (illustrated schematically, in Fig. 6.3.3 with the in-phase sp^3 and $1s$ orbital combination) leads to the experimentally observed structure of the monoanion. (Strangely enough, neutral B_3H_9 has not, as yet, been isolated.)

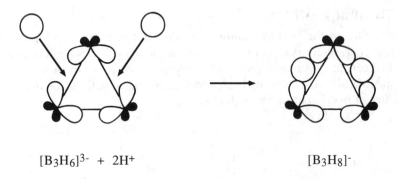

$$[B_3H_6]^{3-} + 2H^+ \qquad\qquad [B_3H_8]^-$$

Fig. 6.3.3 — Protonation (H^+ = empty $1s$ AO) of the B–B bonds, formed by the overlap of sp^3 hybrid orbitals, in the hypothetical $[B_3H_6]^{3-}$ to generate $[B_3H_8]^-$. (Terminal hydrogen atoms have been omitted for clarity.)

The bonding scheme described here for the $[B_3H_8]^-$ anion is a localized one and assumes that the boron atoms are sp^3 hybridized. In this case, such a model appears to give a reasonable bonding picture.

6.4 BORANE CLUSTERS

The bonding in B_4H_{10} could be described by adopting an sp^3 hybridization scheme for the boron atoms, since the structure illustrates 4-coordinate, approximately tetrahedrally sited boron atoms Fig. 6.4.1). However, after B_4H_{10}, life becomes more difficult! Glancing through the borane cluster structures described in Chapter 4, we observe that sp^2 and sp^3 hybridization schemes are certainly *not* always appropriate. How, then, can we rationalize the bonding within these borane cages? Since we should always strive to maintain relatively easy lives for ourselves, a single description which could be applied to *all* the boranes would be more than welcome.

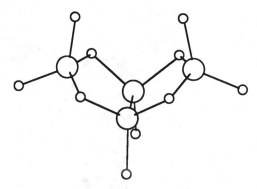

Fig. 6.4.1 — Structure of B_4H_{10} emphasizing 4-coordinate boron atoms.

Borane structures are, as we have seen, derived from a set of closed deltahedra (Fig. 4.1.1), and each polyhedron is *approximately* spherical in shape. Thus, if one, and only one, adjective were sought to describe *all* the cages, 'spherical' would be the most suitable. In the majority of cases, each boron atom has attached to it one terminal hydrogen atom. If we consider both these features as criteria for choosing a suitable hybridization scheme for the boron atom, then *sp* hybridization is most appropriate. While one *sp* hybrid orbital points towards the terminal hydrogen atom, the second is oriented towards the centre of the borane cage (Fig. 6.4.2(a) vs. 6.4.2(b)). This will always be true, irrespective of the local symmetry of the boron atom in the cage. The inward-pointing hybrid orbital is called a *radial* orbital. In addition to the *sp* hybrid orbitals, each boron atom possesses two $2p$ AOs, directed tangentially with respect to the surface of the 'spherical' cluster (Fig. 6.2.4(b)). Thus, after forming a localized 2-centre 2-electron B–H$_{(terminal)}$ bond, each boron atom has available one *radial* and two *tangential* orbitals. These orbitals are called the *frontier orbitals* of the BH fragment, and their generation from the atomic orbitals of a boron and hydrogen atom is illustrated in the MO correlation diagram drawn in Fig. 6.4.3. Note that the frontier orbitals possess only boron character, and are available for binding the fragment to adjacent units in the cluster.

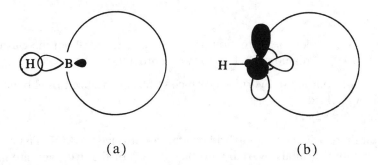

(a)   (b)

Fig. 6.4.2 — Schematic representation of a BH unit on the surface of a 'spherical' cluster: (a) one *sp* hybrid orbital is used for the localized BH interaction; (b) after BH bond formation, a radial *sp* hybrid and two tangential $2p$ AOs remain.

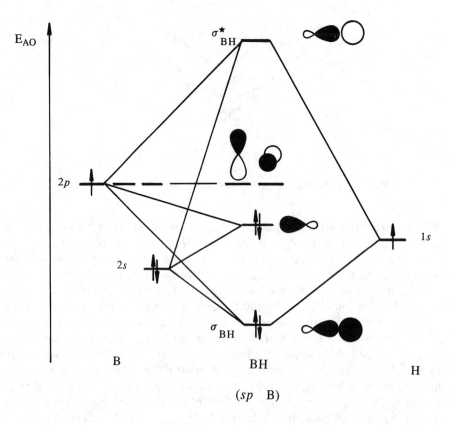

Fig. 6.4.3 — MO correlation diagram for the generation of a BH fragment; note the three frontier orbitals of a BH fragment which are nonbonding with respect to the B–H bond.

Let us now consider the formation of a specific deltahedral borane cluster: the octahedral $[B_6H_6]^{2-}$ dianion. Each boron atom provides:

(1) an outward-pointing sp hybrid with which to form a localized B–H bond
(2) a radial sp hybrid
(3) two tangential $2p$ AOs

and in total there are twelve sp hybrid orbitals and twelve $2p$ AOs. The outward-pointing sp hybrid orbitals overlap with the $1s$ AOs of the hydrogen atoms (Fig. 6.4.4), leaving eighteen boron-based orbitals to combine to form cluster MOs. In Fig. 6.4.4, these cluster orbitals are partitioned into seven bonding and eleven non- and antibonding MOs, the construction of which is as follows.

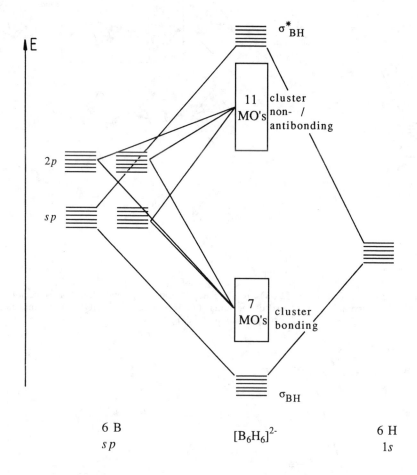

Fig. 6.4.4 — Correlation of boron (*sp* hybridized) and hydrogen orbitals in [B₆H₆]²⁻; a detailed analysis of the cluster MOs is given in Fig. 6.4.6.

There are six radial *sp* hybrid orbitals in the B₆-cluster, and we can therefore construct six LGOs using just these hybrids. The lowest energy LGO is unique (ψ_1 in Fig. 6.4.5), and arises from an in-phase combination of all six radial orbitals. We can then draw three equivalent (degenerate) LGOs (ψ_2–ψ_4), each of which contains a single nodal plane. Finally, a set of three LGOs, each with 2 nodal planes, is constructed. But now a problem appears: *seven* LGOs have been constructed when only six are allowed! Taking a linear combination of two degenerate LGOs solves this crisis and leaves us with ψ_5 and ψ_6 (Fig. 6.4.5). The six *sp* hybrid orbitals are correlated with their associated LGOs on the left-hand side of Fig. 6.4.6.

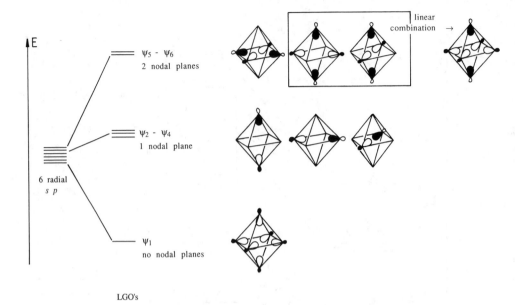

Fig. 6.4.5 — The formation of LGOs from radial *sp* hybrids in an octahedral B_6-cluster; note that a linear combination of two of the initial LGOs is necessary. LGO orbital numbers correspond to those in Fig. 6.4.6.

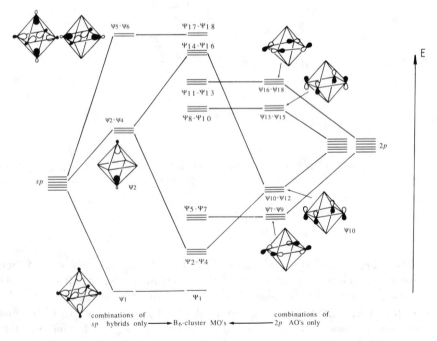

Fig. 6.4.6 — Correlation of the frontier orbitals of the six boron atoms in $[B_6H_6]^{2-}$; the left-hand side of the diagram illustrates the generation of LGOs from *sp* hybrids only for the B_6-cage, whilst combinations of the tangential *2p* AOs are shown on the right-hand side of the figure. Interaction of the two sets of LGOs to generate B_6-cluster MOs is shown in the centre of the diagram.

Consider, now, the tangential orbitals of the B_6-cage. For a given boron atom, the two $2p$ AOs are orthogonal, and combinations of these orbitals generate four triply degenerate groups of LGOs (two bonding, ψ_7–ψ_{12}, and two antibonding, ψ_{13}–ψ_{18}, correlated on the right-hand side of Fig. 6.4.6). The two sets of LGOs originating respectively from the sp and $2p$ orbitals are *not* mutually exclusive. For example, the symmetry of ψ_2 matches that of ψ_{10} and this permits ψ_2 and ψ_{10} to mix, giving bonding and antibonding combinations; similarly, mixing between ψ_3 and ψ_{11} and between ψ_4 and ψ_{12} occurs. These correlations are shown in Fig. 6.4.6. The eighteen MOs of the B_6-cluster separate into seven low and eleven high lying orbitals, and it is this energy distinction which determines how many cluster bonding electrons the octahedral B_6-cage requires. The large energy gap observed in Figs 6.4.4 and 6.4.6 defines the HOMO-LUMO separation of the cluster, and, hence, the total number of bonding MOs and, therefore, of bonding pairs of electrons, is thirteen (six σ_{BH} plus seven B_6-cluster bonding). However, each BH unit provides only 4 electrons (see Fig. 6.4.3) and thus the cluster must exist as a dianion, viz. $[B_6H_6]^{2-}$. The seven cluster bonding MOs (along with appropriate symmetry labels) of $[B_6H_6]^{2-}$ are represented in Fig. 6.4.7.

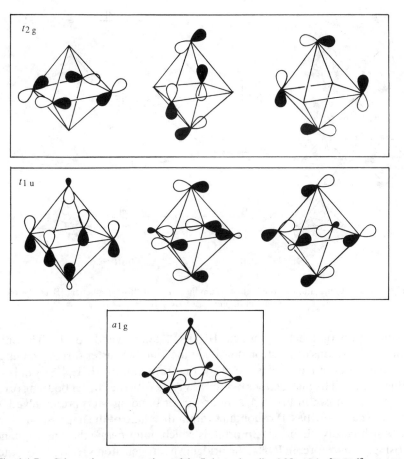

Fig. 6.4.7 — Schematic representations of the 7 cluster bonding MOs of the $[B_6H_6]^{2-}$ dianion.

A molecular orbital analysis analogous to the one above may be applied to any borane cage, closed or open. The structure of pentaborane(9) is related to that of $[B_6H_6]^{2-}$, so let us examine the relationship between the MOs of this neutral borane and those of the closed octahedral cluster. Setting aside the localized terminal B–H bonds, each $BH_{(terminal)}$ unit provides 3 frontier orbitals and 2 valence electrons as before (Fig. 6.4.3). Let us initially generate a hypothetical square-based pyramidal $[B_5H_5]^{4-}$ anion. The frontier orbitals of the 5 BH fragments combine to give 7 cluster bonding MOs and 8 nonbonding/antibonding MOs which are readily constructed by recognizing that the $[B_5H_5]^{4-}$ anion is formally generated from $[B_6H_6]^{2-}$ by the removal of a $[BH]^{2+}$ fragment. Fig. 6.4.8 illustrates what happens to each of the cluster bonding MOs of $[B_6H_6]^{2-}$ as this vertex is removed. One critical feature is that the destruction of the octahedral symmetry causes the triply degenerate (t_{1u} and t_{2g}) MOs to split. However, the most important observation is that the total number of cluster bonding MOs is conserved in going from the closed B_6- to the open B_5-cage.

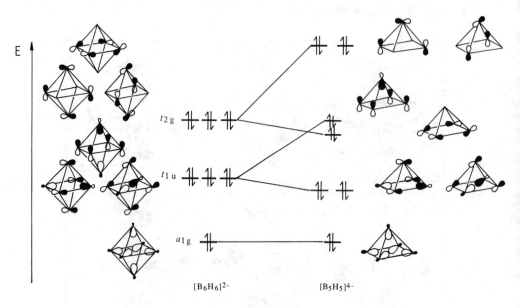

Fig. 6.4.8 — Generation of the 7 cluster bonding MOs of the hypothetical $[B_5H_5]^{4-}$ by the removal of one vertex from $[B_6H_6]^{2-}$.

Protonation of the hypothetical $[B_5H_5]^{4-}$ anion yields B_5H_9. The incoming protons seek regions of electron density in the anionic cluster surface; compare this to the analogous situation described for the formation of $[B_3H_8]^-$ from $[B_3H_6]^{3-}$ (see Fig. 6.3.3). The interactions of the H $1s$ AOs with the cluster bonding orbitals of $[B_5H_5]^{4-}$ are shown in Fig. 6.4.9. Four cluster bonding MOs are involved, one of which is B(basal)–B(basal) σ-bonding within the plane of the open square face, and three which involve B(basal) $2p$ orbitals which point *below* the square plane. Not surprisingly, therefore, in B_5H_9, the bridging hydrogen atoms reside *below* the plane of the four basal boron atoms (see Fig. 4.5.1(a)).

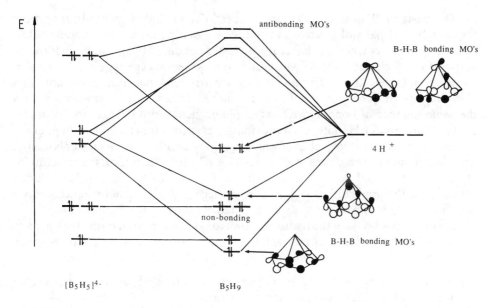

Fig. 6.4.9 — Orbital correlation diagram for the protonation of $[B_5H_5]^{4-}$ to form B_5H_9. Note that only four MOs are influenced by the approach of the H $1s$ AOs.

Two important features emerge from the bonding description above.

(1) The open square-based pyramidal B_5-cage possesses the *same* number of cluster bonding MOs as the closed octahedral B_6-cage. Therefore, the number of bonding molecular orbitals depends not only on the number of atoms but on the geometry of the cluster as well.

(2) The removal of a vertex from the B_6-cage leaves an open face possessing a surfeit of electron density which is readily available for interaction with electrophiles, e.g. protons.

These phenomena extend to borane cages in general and form the basis for Wade's rules and the Polyhedral Skeletal Electron Pair Theory (PSEPT) of cluster bonding.

6.5 WADE'S RULES AND THE POLYHEDRAL SKELETAL ELECTRON PAIR THEORY (PSEPT)

The construction of molecular orbitals for clusters, particularly large and open cages, is not a trivial matter without the aid of computers. Wade's rules and the Polyhedral Skeletal Electron Pair Theory (PSEPT) developed principally by Wade [1], Williams [2], and Mingos [3], provide an empirical set of rules for predicting (or, conversely, for rationalizing) the structures of cluster molecules. The theories are best suited for clusters of main group elements and are, therefore, ideal for the boranes and for boron-rich metalloboranes. The structures of some metal-rich metalloboranes may also be rationalized by using Wade's rules[†], but others (see section 6.8) present us with a challenge.

[†]The term 'Wade's rules' will be used here to refer to cluster electron counting rules generally.

The basis of Wade's rules is that a closed deltahedral cluster with n vertices requires $(n + 1)$ pairs of electrons to completely occupy the *cluster* bonding MOs. Thus, a 6-vertex octahedral cluster has 7 cluster bonding MOs, a dodecahedron (8 vertices) has 9 cluster bonding MOs, and an icosahedral cage requires 13 pairs of cluster bonding electrons. *Throughout, we are assuming that each boron atom in a closed cage possesses one terminal hydrogen atom.* Moreover, any open cage exhibits the same number of bonding MOs as its parent deltahedron. Take, for example, B_5H_9. There are 5 BH units, each providing 2 electrons for cluster bonding, and 4 additional hydrogen atoms, each of which provides one electron for cluster bonding. The total number of electrons available is 14, or 7 pairs. Therefore, the structure of B_5H_9 should be based upon an octahedron, i.e. the 6-vertex deltahedron has 7 cluster MOs. Thus, the predicted structure of the B_5-cage in B_5H_9 is that of an octahedron with one vertex removed, just as we described above (Figs 6.4.8 and 6.4.9).

Choosing positions for the hydrogen atoms on the boron atom framework usually works as follows. After allowing each boron atom to carry *one* terminal hydrogen atom, the placement of additional hydrogen atoms follows 3 rules:

(1) hydrogen atoms bridge edges round the open face which remains after the removal of a vertex from a closed polyhedron
(2) additional terminal hydrogen atoms may be added to any boron atom which is in a site of extremely low connectivity, e.g. in B_4H_{10} (see below)
(3) the symmetry of the boron core cluster should be preserved as far as possible.

Borane clusters may be categorized according to the degree of openness of the cage:

(1) *closo* — closed deltahedral cage, e.g. $[B_6H_6]^{2-}$
(2) *nido* — open cage related to a closed deltahedral cage by the removal of *one* vertex, e.g. B_5H_9
(3) *arachno* — open cage related to a closed deltahedral cage by the removal of *two* vertices, e.g. B_4H_{10}
(4) *hypho* — open cage related to a closed deltahedral cage by the removal of *three* vertices, e.g. $B_5H_9(PMe_3)_2$
(5) *conjuncto* — fused cages, open or closed, e.g. $2,2'-\{B_{10}H_{13}\}_2$.

The following worked examples illustrate how Wade's rules are applied. The first step is always to identify as many {BH} units as possible, and then to write down the residual hydrogen atoms, and the overall charge, if any.

Example 1: B_5H_9

The formula B_5H_9 is reduced to 5 BH + 4 H

Number of valence electrons for cluster bonding:

$$5\ BH\ =\ 10\ e$$
$$4\ H\ =\ 4\ e$$
$$Total\ =\ 14\ e\ =\ 7\ pairs$$

7 pairs of electrons are consistent with a 6-vertex, octahedral parent deltahedron, but since there are only 5 boron atoms in B_5H_9, there is 1 vertex missing:

The 4 *endo*-H atoms are positioned around the open face of the square-based pyramid, and adopt B–H–B bridging positions, retaining the symmetry of the original B_5-cage. The structure of B_5H_9 is therefore a *nido*-cluster:

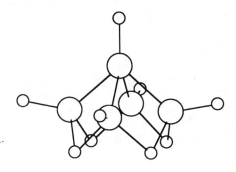

Example 2: B_4H_{10}

The formula B_4H_{10} is reduced to $4\,BH + 6\,H$

Number of valence electrons for cluster bonding:

$$4\,BH = 8\,e$$
$$6\,H = 6\,e$$
$$Total = 14\,e = 7\,pairs$$

7 pairs of electrons are consistent with a 6-vertex, octahedral parent deltahedron, but since there are only 4 boron atoms in B_4H_{10}, there are 2 vertices missing:

Note that the 'butterfly' geometry is preferred over the square, an open cage shape which can also be generated from the octahedron by the removal of 2 vertices.

The 6 *endo*-H atoms are positioned by using the following arguments. Two of the boron atoms, the 'wing-tips' of the butterfly, are of very low connectivity; therefore, each will bear an additional terminal hydrogen atom. The additional 4 hydrogen atoms take up B–H–B bridging positions, retaining the symmetry of the original B_4-cage. The structure of B_4H_{10} is hence an *arachno*-cluster:

Example 3: B_5H_{11}

The formula B_5H_{11} is reduced to $5 BH + 6 H$

Number of valence electrons for cluster bonding:

$$5 BH = 10 e$$
$$6 H = 6 e$$
$$Total = 16 e = 8 pairs$$

8 pairs of electrons are consistent with a 7-vertex, pentagonal bipyramidal parent deltahedron, but since they are only 5 boron atoms, 2 vertices are left vacant:

Several isomers of the B_5-cage are theoretically possible. *In general, the first vertex to be removed is one of highest connectivity and the second is one adjacent to the now vacant site.* As in B_4H_{10}, two of the boron atoms in the B_5-cage are of extremely low connectivity and therefore each carries a second terminal hydrogen atom. Three more *endo*-H atoms bridge the three basal B–B edges of the open face, and the final H atom is unique, as we discussed in section 4.5; it occupies a terminal site attached to the apical boron atom, but simultaneously interacts with the open face of the cluster:

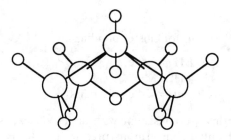

Example 4: B_6H_{10}

The formula B_6H_{10} is reduced to $6\,BH + 4\,H$

Number of valence electrons for cluster bonding:

$$6\,BH = 12\,e$$
$$4\,H = 4\,e$$
$$\text{Total} = 16\,e = 8\,\text{pairs}$$

8 pairs of electrons are consistent with a 7-vertex, pentagonal bipyramidal parent deltahedron, but 1 vertex is unoccupied as only 6 boron atoms are available:

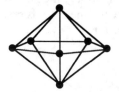 $\rightarrow$

The pyramidal B_6-cage is generated by removing a vertex of high connectivity from the parent deltahedron. The *endo*-H atoms bridge the B–B edges of the open face. However, there is clearly a problem! There are 5 edges and only 4 *endo*-H atoms; hence the 5-fold symmetry of the cage cannot be retained in a static structure. It is therefore reasonable that, in solution, the 4 bridging hydrogen atoms of B_6H_{10} are involved in a fluxional process, exchanging bridging sites around the base of the pentagonal pyramid. The static structure of B_6H_{10} is a *nido*-cage:

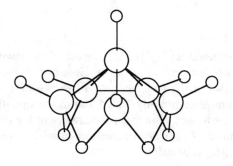

Example 5: $B_{10}H_{14}$

Number of valence electrons for cluster bonding:

$$10\ BH = 20\ e$$
$$4\ H = 4\ e$$
$$Total = 24\ e = 12\ pairs$$

12 pairs of electrons are consistent with an 11-vertex, octadecahedral parent deltahedron, but with only 10 boron atoms present, 1 vertex is unoccupied. The atom of highest connectivity is removed:

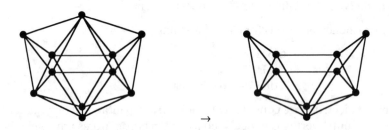

$\rightarrow$

The open face accommodates the 4 *endo*-hydrogen atoms, the sites for the bridges being chosen so as to maintain as high a symmetry for the cage as possible.

The structure of $B_{10}H_{14}$ is therefore a *nido*-cluster:

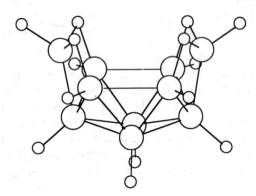

Example 6: $[B_{10}H_{14}]^{2-}$

Reduction of $B_{10}H_{14}$ generates $[B_{10}H_{14}]^{2-}$. The addition of two electrons means that the new B_{10}-cage will be derived from a 12-vertex rather than an 11-vertex deltahedron. Two views of the icosahedron are shown on the left below; the first orientation is the one most commonly drawn, but the second illustrates more clearly that the icosahedron possesses similar characteristics to the octadecahedron shown in the previous example. Removal of 2 vertices provides the same B_{10}-cage upon which the neutral $B_{10}H_{14}$ is based:

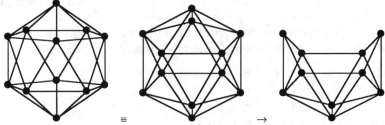

Thus, upon reduction, we predict that the core structure of $B_{10}H_{14}$ changes little, although, in fact, the locations of the *endo*-H atoms do alter:

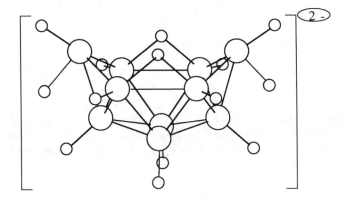

The difference in *endo*-hydrogen atom positions in going from $B_{10}H_{14}$ to $[B_{10}H_{14}]^{2-}$ is not readily predicted by using Wade's rules.

6.6   EXTENSION OF WADE'S RULES TO *CONJUNCTO*-CLUSTERS

The bonding in borane clusters which are coupled through a direct B–B bond is simply rationalized by considering the individual cages, and acknowledging that one cage is acting as an *exo*-ligand to the other. For example, in $\{B_5H_8\}_2$ (see Figs 4.13.1 and 4.13.2), each B_5H_8-unit may be considered to be a free radical. Thus, it is irrelevant whether this combines with $H^\bullet$ or $B_5H_8^\bullet$ to give B_5H_9 or $\{B_5H_8\}_2$ respectively. Hence, Wade's rules can be applied to each sub-unit of the coupled system, but not readily to the system as a whole.

Application of Wade's rules to *conjuncto*-cages involving shared boron atoms requires care. Although the valence electrons of the shared atoms are delocalized over both sub-units of the coupled cage, we must not allow these electrons to be counted twice. Consider the structure of $B_{12}H_{16}$ (Fig. 4.13.6) which comprises B_8- and B_6-cages coupled via a shared B–B bond. The boron atoms at the point of fusion use *all three of their valence electrons for cluster bonding*. The structure may be reduced to a B_6H_7-unit and a B_8H_9-cage. However, since the two cages share two boron atoms, only *one* shared atom 'belongs' to a given sub-unit. From the limited data available, the sub-clusters appear to be derived from parent *isocloso*-clusters. An *isocloso*-cluster is one possessing the *same* number of vertices as pairs of bonding electrons. The electron distribution for cluster bonding in $B_{12}H_{16}$ is thus

B_6H_7 sub-unit: 4 BH = 8 e
                       1 B = 3 e
                       3 H = 3 e

Total cluster electrons = 14 e = 7 pairs

Parent *isocloso*-deltahedron = 7-vertex pentagonal bipyramid
1 vertex removed   = pentagonal pyramid

B_8H_9 sub-unit: 6 BH = 12 e
                       1 B = 3 e
                       3 H = 3 e

Total cluster electrons = 18 e = 9 pairs

Parent *isocloso*-deltahedron = 9 vertex tricapped trigonal prism
1 vertex removed   = bicapped trigonal prism

The predicted structure should therefore be the combination of a pentagonal pyramidal cage and a bicapped trigonal prismatic cage. Careful inspection of Fig. 4.13.6 reveals these features (*hint!* the pentagonal pyramid is upside-down on the right-hand side of the diagram).

6.7   VALENCE BOND MODELS OF BONDING IN BORANES (*styx* RULES)

The structures of boron hydrides may be rationalized in terms of the simple valence bond model developed by Lipscomb [4]. The method is readily applied to small molecules, but becomes more tedious for large cages. The basis of the concept is to construct a network of 2-centre B–H and B–B, and 3-centre B–H–B and B–B–B bonds which will satisfy the valence requirements of both the boron and the hydrogen atoms. The 3-centre B–B–B bonds fall into two classes, open and closed:

open interaction = B–B–B   closed interaction =

For a borane, B_pH_{p+q} there are p BH units and an additional q H atoms. The topology of the borane cage is determined by solving the following simultaneous equations:

$$s + x = q$$
$$s + t = p$$
$$t + y = p - q/2$$

where
   s = number of 3-centre 2-electron B–H–B bridges
   t = number of 3-centre 2-electron B–B–B (open or closed) interactions
   y = number of 2-centre 2-electron B–B bonds
   x = number of BH_2 units.

For the borane under study, values of p and q are defined: e.g. for pentaborane(9), $p = 5$ and $q = 4$. Substitution for p and q in the above equations with trial values of s $(0 \leqslant s \leqslant p)$ permits solutions for t, x and y to be determined. Negative values are invalid (see the definitions of t, x and y), but positive values represent possible topologies for the borane cage. A worked example is given below.

   B_5H_9: $p = 5$ $q = 4$

      Let $(0 \leqslant s \leqslant 5)$

   Solve the simultaneous equations

   $$s + x = q \tag{i}$$
   $$s + t = p \tag{ii}$$
   $$t + y = p - q/2 \tag{iii}$$

Let $s = 0$, Therefore in (i) $x = 4$, and in (ii) $t = 5$, but from (iii) $y = -2$. Therefore solution is not real.

Let $s = 1$, Therefore in (i) $x = 3$, and in (ii) $t = 4$, but from (iii) $y = -1$. Therefore solution is not real.

Let $s = 2$, Therefore in (i) $x = 2$, (ii) $t = 3$, and from (iii) $y = 0$. Therefore solution is real.

Let $s = 3$, Therefore in (i) $x = 1$, in (ii) $t = 2$, and from (iii) $y = 1$. Therefore solution is real.

Let $s = 4$, Therefore in (i) $x = 0$, in (ii) $t = 1$, and from (iii) $y = 2$. Therefore solution is real.

Let $s = 5$, Therefore in (i) $x = -1$, in (ii) $t = 0$, and from (iii) $y = 3$. Therefore solution is not real.

The only possible topologies for B_5H_9 correspond to *styx* combinations of (2302), (3211), or (4120). Recalling the definitions of *styx* allows the cage networks illustrated in Fig. 6.7.1 to be constructed. After comparison with the known structure of pentaborane(9), it is apparent that form (4120) is favoured, although it is not easy to argue in advance which of the three predicted networks would be preferred by the molecule.

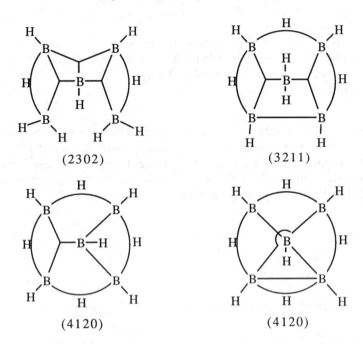

(2302)   (3211)

(4120)   (4120)

Fig. 6.7.1 — *styx* topologies for B_5H_9. There are two ways of constructing the (4120) network.

6.8   THE INCORPORATION OF METAL FRAGMENTS: THE ISOLOBAL PRINCIPLE

In Chapter 5, we classified metalloboranes according to their structural type. Not surprisingly, descriptions of the bonding within these clusters follow the same classifications.

A metal fragment which functions as an *exo*-ligand effectively replaces an *exo*-hydrogen in a borane cage, and the bonding of the metal atom to the borane cluster is via the localized 2-centre 2-electron bond. For an *endo*-metal fragment in a B–M–B bridging site, a 3-centre 2-electron bonding mode is a valid description. A suitable hybrid orbital (σ-symmetry) on the metal fragment must be available, for example, in going from a B–H–B bridge to a B–($AuPR_3$)–B bridge (Fig. 6.8.1), the 1s orbital of the proton is replaced by an spd_{z^2} hybrid (which happens to possess little d-character) from the gold(I) phosphine fragment.

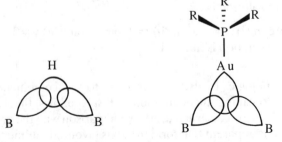

Fig. 6.8.1 — Bonding mode for an *endo*-metal fragment: comparison of the 3-centre 2-electron B–H–B bond with the B–($AuPR_3$)–B, (R = alkyl, aryl) bridge bond.

Consider, now, the incorporation of a metal fragment as a *cluster unit* in a borane cage. In general, the structure of the metalloborane bears a striking resemblance to the structure of a binary boron hydride. For example, the structures of B_5H_9, $Fe(CO)_3B_4H_8$ [5], $Fe_2(CO)_6B_3H_7$ [6], and $Ru_3(CO)_9B_2H_6$ [7] are related as indicated in Fig. 6.8.2. Formally, substitution by an $Fe(CO)_3$ or $Ru(CO)_3$ for a BH fragment has occurred. The cluster retains its square-based pyramidal framework, and four bridging hydrogen atoms are located around the open face in each cage. Of course, a metal atom is larger than a boron atom and usually bears ligands which are more sterically demanding than a terminal hydrogen atom, but distortion due to steric factors apart, the pyramidal geometry is unchanged.

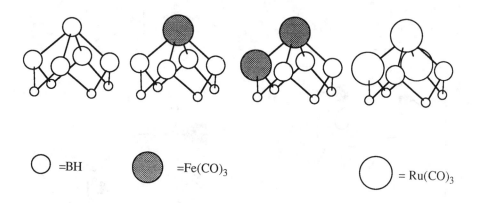

Fig. 6.8.2 — Schematic representations of the core structures of B_5H_9, $Fe(CO)_3B_4H_8$, $Fe_2(CO)_6B_3H_7$ and $Ru_3(CO)_9B_2H_6$ illustrating the retention of cluster geometry.

Structural relationships between boranes and metalloboranes are not coincidental! Formal metal-to-boron substitution occurs if the bonding capabilities of the metal group successfully mimic those of the missing borane fragment. Two cluster fragments are said to be *isolobal* if they exhibit the same number of frontier orbitals and the same number of valence electrons, and if the frontier orbitals of each fragment have matching symmetries. A simple example is the replacement of a $[BH]^-$ unit by CH; the two fragments are isoelectronic and clearly exhibit the same frontier orbitals. (Replacement of $[BH]^-$ by CH units in boron hydride cages generates carborane clusters; see Chapter 7.) Let us compare the MOs of a BH unit (Fig. 6.4.3) with those of a C_{3v} (M(CO)$_3$ (M = Fe, Ru, Os) fragment (Fig. 6.8.3). All but two of the occupied MOs of the M(CO)$_3$ fragment categorize, in order of increasing energy, as either C–O bonding, M–CO bonding, or metal nonbonding. These MOs are all localized *within* the M(CO)$_3$ fragment. In contrast, the HOMO of the M(CO)$_3$ fragment comprises a degenerate set (*e* symmetry) of *outward-pointing* p_xd_{xz} and p_yd_{yz} hybrid orbitals.[†] The LUMO is an outward-pointing $sp_zd_{z^2}$ hybrid.

[†] The choice of axes is arbitrary.

These 3 orbitals, containing 2 electrons, constitute the frontier MOs of the C_{3v} M(CO)$_3$ fragment. They compare favourably with the frontier MOs of the BH fragment since their symmetries, so far as for cluster bonding is concerned, are clearly equivalent; the relative ordering of the frontier MOs is unimportant. Consequently, any cluster which can accommodate a BH unit should, in principle, be able to accommodate a C_{3v} M(CO)$_3$ (M = Fe, Ru, Os) fragment in its place. (As one might expect, actually synthesizing the compounds is not usually a question of simple fragment substitution!)

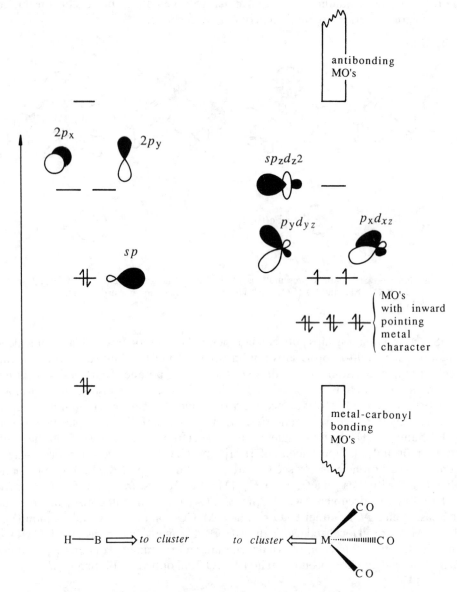

Fig. 6.8.3 — A comparison of the frontier orbitals of a C_{3v} M(CO)$_3$ (M = Fe, Ru, Os) fragment with those of a BH unit; energy scales and choice of axes are arbitrary.

The isolobal principle allows Wade's rules to be extended beyond binary boron hydrides. In the examples above, each $Fe(CO)_3$ or $Ru(CO)_3$ fragment provides 2 cluster bonding electrons, just as does a BH unit. Thus, replacement of BH by either metal fragment has no effect on the overall number of cluster bonding electrons. Selected worked examples are given below.

Example 1: $Fe_2(CO)_6B_3H_7$

The formula is reduced to $2\,Fe(CO)_3$
                                    $3\,BH$
                                    $4\,H$

Number of valence electrons for cluster bonding:

$$2\,Fe(CO)_3 \ = \ 4\,e$$
$$3\,BH \ = \ 6\,e$$
$$4\,H \ = \ 4\,e$$
$$\text{Total} \ = \ 14\,e = 7\ \text{pairs}$$

7 pairs of electrons are consistent with a 6-vertex, octahedral parent deltahedron. There are 5 cluster core atoms (2Fe + 3B) and hence, 1 vertex of the octahedron remains vacant:

   →

The *endo*-H atoms bridge the open square face. The question now is one of possible cluster atom isomerism. Three isomers are feasible as we saw in Fig. 5.7.1, and Wade's rules are unable to distinguish between them. Chemical intuition may suggest that one isomer is preferred, and this can be supported by spectroscopic evidence, but a crystallographic study is the only sure way of solving the dilemma.

Example 2: $Ru_3(CO)_9BH_5$

The formula is reduced to $3\,Ru(CO)_3$
                                    $1\,BH$
                                    $4\,H$

Number of valence electrons for cluster bonding:

$$3\,Ru(CO)_3 \ = \ 6\,e$$
$$1\,BH \ = \ 2\,e$$
$$4\,H \ = \ 4\,e$$
$$\text{Total} \ = \ 12\,e = 6\ \text{pairs}$$

6 pairs of electrons are consistent with a 5-vertex, trigonal bipyramidal parent deltahedron. There are 4 cluster core atoms (3Ru + 1B) and hence, 1 vertex remains vacant.

There are several ways of arranging the *endo*-H atoms so that they bridge edges of the tetrahedral Ru_3B-core and, in practice, two isomers of $Ru_3(CO)_9BH_5$ have been spectroscopically characterized [8]:

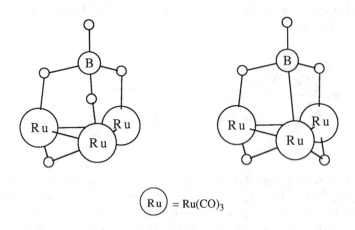

$$\boxed{Ru} = Ru(CO)_3$$

Any metal fragment, transition or main group, exhibits a characteristic set of frontier orbitals, and a characteristic number of valence electrons is available for cluster bonding. It is important to remember that *the nature of the frontier MOs depends upon the geometry of the ligands attached to the metal atom*. Numbers of electrons provided by some common cluster fragments are listed in Table 6.8.1, and these electron counts are calculated assuming that each transition metal atom requires an 18-electron configuration. Not all transition metals obey the Eighteen-electron Rule, and, of the metals listed in Table 6.8.1, palladium and platinum are often stable in a 16-electron configuration. This difference allows *two more* electrons to be released from localized Pd- or Pt-ligand bonding MOs to cluster bonding MOs, and hence the electron count for a fragment involving Pd and Pt requires appropriate adjustment. In Table 6.8.1, note that a negative value implies that the transition metal unit withdraws electron density from the cluster, while in a fragment such as

$Fe(CO)_2$, all three of the frontier MOs are empty. Other electron counts for fragments may be determined from the equation

$$\text{Number of electrons occupying frontier MOs} = v + x - 12$$

where

      v = valence shell electrons of the metal atom in zero oxidation state
      x = valence electrons donated by *exo*-ligands

For example, in a $[(\eta^5\text{-Cp})Fe(CO)]$ cluster fragment, the Fe atom has an electronic configuration of $4s^2 3d^6$; hence, $v = 8$. The carbonyl ligand provides 2 electrons to the Fe–CO bonding while the η^5-Cp ligand provides 5 π-electrons; hence $x = 7$. The $[(\eta^5\text{-Cp})Fe(CO)]$ unit therefore contributes 3 electrons to a cluster.

Table 6.8.1 — Number of electrons contributed by some transition metal fragments to cluster bonding

Metal	Cluster fragment			
	ML_2 L = CO, PR$_3$ or other 2 e donor	ML_3 L = CO, PR$_3$ or other 2 e donor	$M(\eta^5\text{-Cp})$	$M(\eta^6\text{-Bz})$
Cr,Mo,W	− 2	0	− 1	0
Mn,Tc,Re	− 1	1	0	1
Fe,Ru,Os	0	2	1	2
Co,Rh,Ir	1	3	2	3
Ni,Pd,Pt	2	4	3	4

Bz = π-benzene.

    Electron counts for main group, metal and non-metal, cluster units are listed in Table 6.8.2. Each number is derived from the equation

$$\text{Number of electrons occupying frontier MOs} = v + x - 2$$

where

      v = valence shell electrons of the main group element
      x = valence electrons donated by *exo*-ligands

For a main group element, it is common for the *exo*-ligand to be a single hydrogen atom, in which case $x = 1$. A naked (i.e. having no *exo*-ligands) element usually exhibits an *exo*-lone pair of electrons, but may provide all its valence electrons to the cluster. An experimentally determined cluster geometry should distinguish between these two states. If the atom is located on the surface of the cluster it will possess an *exo*-lone pair, but if it is surrounded by other atoms (i.e. it is interstitially bonded), it will make use of all its valence electrons. Examples of interstitial atoms are provided in section 6.9.

Table 6.8.2 — Number of electrons contributed by some main group fragments to cluster bonding

Group containing element, X	Cluster fragment		
	X assuming *exo*-lone pair	XH	ML L = 2H, CO, PR$_3$ or other 2 e donor
I	not applicable	0	1
II	0	1	2
III	1	2	3
IV	2	3	4
V	3	4	5
VI	4	5	6

The following examples illustrate how electron counts from Tables 6.8.1 and 6.8.2 may be used in conjunction with Wade's rules to predict metalloborane cluster core geometries. Remember that the orientation of *exo*-ligands *cannot* be rationalized by this method. The only definitive way of obtaining this information is via a crystallographic structural determination.

Example 1:   Ir(PPh$_3$)$_2$(CO)B$_5$H$_8$ [9]

The formula is reduced to   Ir(CO)(PPh$_3$)$_2$
                                     5 BH
                                     3 H

Number of valence electrons for cluster bonding:

$$Ir(CO)(PPh_3)_2 = 3\,e$$
$$5\,BH = 10\,e$$
$$3\,H = 3\,e$$
$$Total = 16\,e = 8\,pairs$$

8 pairs of electrons → 7-vertex pentagonal bipyramid
6 cluster core atoms (Ir + 5B) → *nido*-cluster

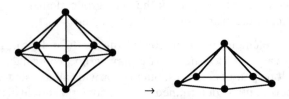

The iridium group occupies a basal site (see Fig. 5.7.2(a)), and this permits the 5 boron atoms to retain a triangular-faced framework.

The structure of Mn(CO)$_3$B$_5$H$_{10}$ has an analogous geometry [10].

Example 2: $Co(\eta^5\text{-}Cp)B_4H_8$ [11]

The formula is reduced to $Co(\eta^5\text{-}Cp)$
                                 4 BH
                                 4 H

Number of valence electrons for cluster bonding:

$$Co(\eta^5\text{-}Cp) = 2\,e$$
$$4\,BH = 8\,e$$
$$4\,H = 4\,e$$
$$Total = 14\,e = 7\ pairs$$

7 pairs of electrons $\rightarrow$ 6-vertex octahedron
5 cluster core atoms (Co + 4B) $\rightarrow$ *nido*-cluster, square pyramid.

The 4-*endo*-H atoms bridge the edges of the open face (cf. B_5H_9). Both geometrical isomers are observed:

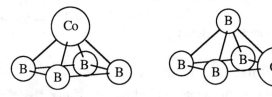

Example 3: $[Ni(\eta^5\text{-}Cp)B_{11}H_{11}]^-$ [12]

The formula is reduced to $Ni(\eta^5\text{-}Cp)$
                                 11 BH
                                 1^- charge

Number of valence electrons for cluster bonding:

$$Ni(\eta^5\text{-}Cp) = 3\,e$$
$$11\,BH = 22\,e$$
$$1^-\ charge = 1\,e$$
$$Total = 26\,e = 13\ pairs$$

13 pairs of electrons $\rightarrow$ 12-vertex icosahedron
12 cluster core atoms (Ni + 11B) $\rightarrow$ *closo*-cluster

All icosahedral sites are equivalent and there are no *endo*-hydrogen atoms. Therefore the structure of the NiB_{11}-core is proposed to be

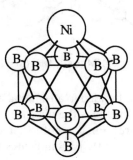

Example 4: $Fe(\eta^6\text{-}C_6H_5Me)SB_{10}H_{10}$ [13]

The $\eta^6\text{-}C_6H_5Me$ ligand is equivalent to a 6π-electron benzene ligand.

The formula is reduced to $\quad Fe(\eta^6\text{-}C_6H_5Me)$
$$S$$
$$10\ BH$$

Number of valence electrons for cluster bonding:

$$Fe(\eta^6\text{-}C_6H_5Me) = 2\ e$$
$$S = 4\ e$$
$$10\ BH = 20\ e$$
$$Total = 26\ e = 13\ pairs$$

13 pairs of electrons $\rightarrow$ 12-vertex icosahedron
12 cluster core atoms (Fe, S + 10B) $\rightarrow$ *closo*-cluster

Isomerism is obviously a possibility, crystallographic results indicate the presence of only one isomer in the solid state, one in which the S and Fe atoms occupy adjacent positions in the icosahedral skeleton.

6.9 METALLOBORIDE CLUSTERS

Examples of metal-rich metallobor*ane* clusters include $H_2\{(\eta^5\text{-}Cp)Co\}_4B_2H_2$, $Fe_2(CO)_6B_2H_6$, $Fe_3(CO)_9BH_5$, $Ru_3(CO)_9BH_5$ and $Ru_3(CO)_9B_2H_6$, the structures of which may be rationalized in terms of Wade's rules. Each boron atom carries a *terminal* hydrogen ligand. On the other hand, some metal-rich, boron-containing clusters are more reasonably described as metallo*borides*. The new definition is applicable whenever we have a boron atom which does not carry a terminal hydrogen atom. An interesting example, and one which bridges the metalloborane-boride categories, is $HM_4(CO)_{12}BH_2$ (M = Fe,Ru)

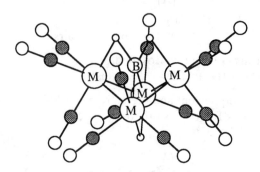

Fig. 6.9.1 — Structure of $HM_4(CO)_{12}BH_2$ (M = Fe [14] and Ru [15].

(Fig. 6.9.1). The boron atom is bonded to all 4 metal atoms, and 3 *endo*-H atoms are associated with the cluster. How does the absence of a $BH_{(terminal)}$ unit affect the rationalization of the structure? Firstly, note that the crystallographically determined Fe_4B-geometry of $HFe_4(CO)_{12}BH_2$ indicates that the boron atom is intimately associated with the metal atoms; it resides *within* the Fe_4-butterfly and is classed, therefore, as an interstitial atom. Application of Wade's rules is as follows:

The formula $HM_4(CO)_{12}BH_2$ (M = Fe, Ru) is reduced to   4 M(CO)$_3$
$$1\,B$$
$$3\,H$$

Number of valence electrons for cluster bonding:

$$4\,M(CO)_3 = 8\,e$$
$$1\,B = 3\,e$$
$$3\,H = 3\,e$$
$$Total = 14\,e = 7\,pairs$$

Note that, since the B atom is interstitially sited, it contributes *all* its valence electrons to cluster bonding. A consequence is that the B atom is not included in the vertex count, since it resides *inside* the cluster:

7 pairs of electrons → 6-vertex, octahedron
4 cluster core atoms (4Fe or 4Ru) → *arachno*-cluster

The metalloboride cluster anion, $[Fe_4Rh_2(CO)_{16}B]^-$, exhibits a *closo*-octahedral core with the boron atom buried neatly inside the metal framework. Two isomers are possible (Rh atoms adjacent or apart), and both are observed [16]. Again, structural rationalization using Wade's rules succeeds. In contrast, the metalloboride $Fe_4(CO)_{12}(AuPPh_3)_3B$ [17], Fig. 6.9.2) presents a problem in terms of a bonding rationale. The seven metal atoms are not arranged in a regular 7-vertex deltahedron, nor do they conform to a fragment of any recognizable polyhedron! The problem arises because of the inclusion of Au(PR$_3$) units. We noted in section 5.5 and in Fig. 6.8.1 that a gold(I) phosphine fragment may mimic the bonding mode of an *endo*-hydrogen atom. So, in fact, Au(PR$_3$) moieties are best described as *endo*-ligands, each being a 1-electron donor in terms of Wade's approach. However, whereas *endo*-H atoms tend to be nonbonded with respect to one another, the same restriction does not apply to *endo*-Au(PR$_3$) groups. The structure of $Fe_4(CO)_{12}(AuPPh_3)_3B$ is related to that of $HFe_4(CO)_{12}BH_2$, but unlike hydrogen atoms, the gold atoms aggregate together in the cluster skeleton. It is impossible to predict with confidence the irregular core geometry for the $Fe_4(CO)_{12}(AuPPh_3)_3B$ cluster. Use of Wade's rules is limited to the expectation of an Fe_4-butterfly core encapsulating an interstitial boron atom, and which has associated with it 3 *endo*-AuPPh$_3$ fragments.

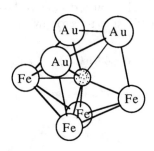
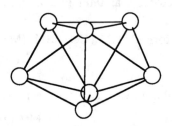

Fig. 6.9.2 — The crystallographically determined Fe_4Au_3B core-structure of $Fe_4(CO)_{12}(AuPPh_3)_3B$ compared with a regular 7-vertex deltahedron.

REFERENCES

[1] K. Wade, *Chem. in Britain*, 1975, **11**, 177; *Adv. Inorg. Chem. Radiochem.*, 1976, **18**, 1.

[2] R. E. Williams, *Adv. Inorg. Chem. Radiochem.*, 1976, **18**, 67.

[3] D. M. P. Mingos, *Nature*, 1972, **236**, 99.

[4] W. N. Lipscomb, *Boron Hydrides*, Benjamin, New York, 1963.

[5] N. N. Greenwood, C. G. Savory, R. N. Grimes, L. G. Sneddon, A. Davison and S. S. Wreford, *J. Chem. Soc., Chem. Commun.*, 1974, 718.

[6] E. L. Anderson, K. J. Haller and T. P. Fehlner, *J. Am. Chem. Soc.*, 1979, **101**, 4390.

[7] A. K. Chipperfield and C. E. Housecroft, unpublished results.

[8] A. K. Chipperfield and C. E. Housecroft, *J. Organomet. Chem.*, 1988, **349**, C17.

[9] N. N. Greenwood, J. D. Kennedy, W. S. McDonald, D. Reed and J. Staves, *J. Chem. Soc., Dalton Trans.*, 1979, 117.

[10] M. B. Fischer and D. F. Gaines, *Inorg. Chem.*, 1979, **18**, 3200.

[11] V. R. Miller and R. N. Grimes, *J. Am. Chem. Soc.*, 1973, **95**, 5078.

[12] B. P. Sullivan, R. N. Leyden and M. F. Hawthorne, *J. Am. Chem. Soc.*, 1975, **98**, 455.

[13] S. O. Kang, P. J. Carroll and L. G. Sneddon, *Organometallics*, 1988, **7**, 772.

[14] T. P. Fehlner, C. E. Housecroft, W. R. Scheidt and K. S. Wong, *Organometallics*, 1983, **2**, 825.

[15] A. K. Chipperfield, C. E. Housecroft and A. L. Rheingold, *Organometallics*, 1990, in press.

[16] T. P. Fehlner, R. Khattar, J. Puga and A. L. Rheingold, *J. Am. Chem. Soc.*, 1989, **111**, 1877.

[17] C. E. Housecroft, A. L. Rheingold and M. S. Shongwe, *J. Chem. Soc. Chem. Commun.*, 1988, 965.

7

Reactivity of boranes and metalloboranes

7.1 THE CHEMISTRY OF SMALL, OPEN-CAGE BORANES

In this section, the definition of a 'small' borane is one with $\leqslant 6$ boron atoms. The simple Lewis acid–base chemistry which is associated with BH_3 and B_2H_6 was described in detail in Chapter 2. Diborane(6) is the precursor to a wealth of higher boranes and derivatives thereof. It is a gas at room temperature (m.p. $= -164.9°C$; b.p. $= -92.6°C$) and hydrolyses readily; its combustion is highly exothermic:

$$B_2H_6 + 6H_2O \rightarrow 2B(OH)_3 + 6H_2$$
$$\text{boric acid}$$

$$B_2H_6 + 3O_2 \rightarrow B_2O_3 + 3H_2O \qquad \Delta H = -2138 \text{ kJ mol}^{-1}$$

The driving force for this reactivity is the strength of a B–O bond relative to a B–H bond. The 3-coordinate boron atom in either $B(OH)_3$ or B_2O_3 is stabilized by π-bonding (refer to Fig. 2.1.8), something which can never be achieved in a B–H bond. The value of the standard heat of formation, $\Delta H_f^{\ominus}$, for diborane is $+36 \text{ kJ mol}^{-1}$, and $\Delta G_f^{\ominus}$ is $+87 \text{ kJ mol}^{-1}$, indicating that the compound is thermodynamically unstable with respect to its constituent elements. However, at ambient temperature and pressure, diborane is kinetically stable, although, when heated, decomposition does occur. Controlled pyrolysis is an effective method of synthesizing higher boranes. Mechanistically, these reactions are far from simple; for instance, a higher boron hydride, e.g. B_5H_{11}, once formed during the pyrolysis of B_2H_6, will undergo decomposition. Changing the reaction conditions influences the product distribution. For example, 90% of the diborane is converted to B_4H_{10} if a hot–cold reactor with an inner surface temperature of 100°C and outer surface at $-80°C$ is used. On the other hand, 80% conversion to B_5H_9 may be achieved if B_2H_6 and H_2, in a ratio 1:5 by volume, are passed through a multistage reactor at 240°C.

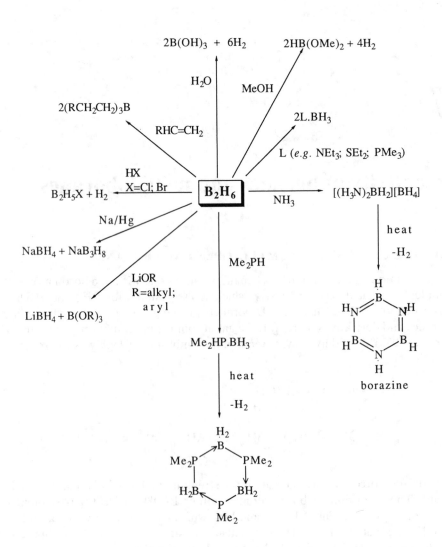

Fig. 7.1.1   Some reactions of B_2H_6.

Some typical reactions of diborane(6) are summarized in Fig. 7.1.1. The elimination of H_2 is a common driving force in the chemistry of boranes. If terminal hydrogen atoms are present, adduct formation may be the first stage in the formation of heteronuclear inorganic rings. The elimination of molecular hydrogen obviously ceases when there is no longer a pair of terminal hydrogen atoms present on adjacent heteroatoms; in Fig. 7.1.1, compare the ring closure reactions which lead to the formation of $[HBNH]_3$ and $[H_2BPMe_2]_3$. Exchange reactions are also important and provide a means of synthesizing substituted diborane. However, if bonds between the substituent and boron are stablized by π-contributions, the dimeric structure of the borane may be lost (e.g. reaction with LiOR or with MeOH). Substitution at the bridging position may be favoured, for example in the reaction of B_2H_6 with SH^- ion:

General reactions of diborane may be extended to other small, open-cage boron hydrides such as *arachno*-B_4H_{10}, *nido*-B_5H_9, *arachno*-B_5H_{11}, *nido*-B_6H_{10} and *arachno*-B_6H_{12}. Deprotonation is now a viable proposition and is, in fact, an important synthetic step towards the formation of metalloborane compounds; a borane anion is a ligand capable of interaction with a metal fragment. Some reactions of tetraborane(10) are represented in Fig. 7.1.2. Attack by Lewis bases is, not surprisingly, a characteristic feature of the chemistry of B_4H_{10}. However, the mechanism of nucleophilic attack is dependent upon the base. With ammonia, 2 moles of the base attack the same B(2) atom (see Fig. 7.1.2 for nomenclature). The reaction is analogous to the asymmetrical cleavage of diborane, and the product is a 1:1 electrolyte (Fig. 7.1.3). With the more sterically demanding trimethylamine, one mole of amine attacks B(1) and one, B(2) (Fig. 7.1.3). This mechanism is analogous to the symmetrical cleavage of diborane and leads to the covalently bonded adducts, $Me_3N.BH_3$ and $Me_3N.B_3H_7$. Reaction with secondary amines results in the cleavage of B_4H_{10}, elimination of hydrogen and the formation of an amido bridge (Fig. 7.1.2).

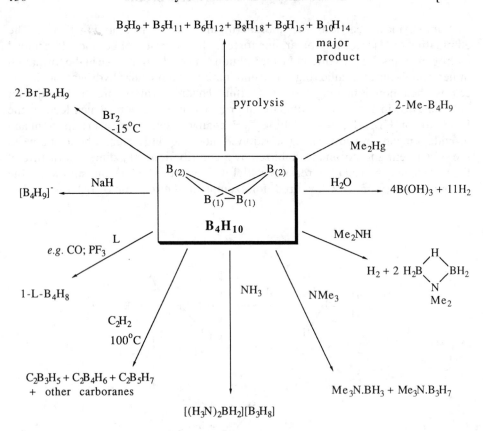

Fig. 7.1.2 — Some reactions of B_4H_{10}.

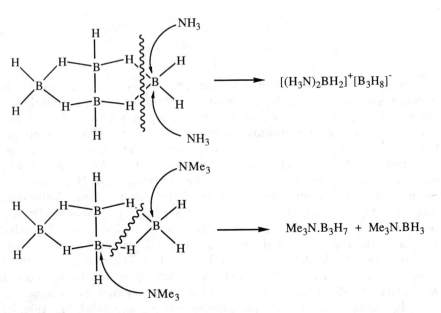

Fig. 7.1.3 — Cleavage of B_4H_{10} by amines.

Since a C atom is isoelectronic with B^-, a series of cluster compounds may be derived, each member of which is related to a boron hydride by the formal replacement of BH^- or BH_2 fragments by CH units. A general synthetic route to these *carborane* clusters is the reaction of boranes with acetylenes, and is illustrated by the reaction of B_4H_{10} with C_2H_2. Three carboranes so formed are $C_2B_3H_5$, $C_2B_4H_6$ and $C_2B_5H_7$, the structures of which may be rationalized by Wade's approach. For example:

$C_2B_3H_5$ reduces to the fragments: $2CH = 2BH^- = 2 \times 3e$

$$3BH = 3 \times 2e$$

Therefore total electrons for cluster bonding $= 12e = 6$ pairs

Therefore parent closed polyhedron is a 5-vertex, trigonal bipyramid

There are 5 cluster atoms $(2C + 3B)$, hence $C_2B_3H_5$ is a *closo*-carborane.

As with metalloboranes, the question of isomerism arises when we consider the arrangement of the cage atoms in a carborane. Three possible isomers are available for $C_2B_3H_5$ and these are illustrated in Fig. 7.1.4. The thermodynamically favoured isomer is the one which maximizes the number of B–C bonds, i.e. the isomer in which the two carbon atoms are remote from one another. Similarly for $C_2B_4H_6$ and $C_2B_5H_7$, *closo*-cage structures are predicted (Fig. 7.1.4), and an isomer having non-adjacent carbon atoms is thermodynamically preferred. Where there is a choice, the carbon atoms tend to avoid the sites of high electron density, e.g. apical positions.

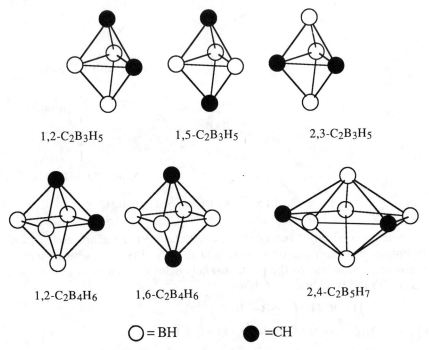

1,2-C2B3H5   1,5-C2B3H5   2,3-C2B3H5

1,2-C2B4H6   1,6-C2B4H6   2,4-C2B5H7

○ = BH   ● = CH

Fig. 7.1.4 — *Closo*-carborane clusters with possible isomers illustrated for $C_2B_3H_5$ and $C_2B_4H_6$.

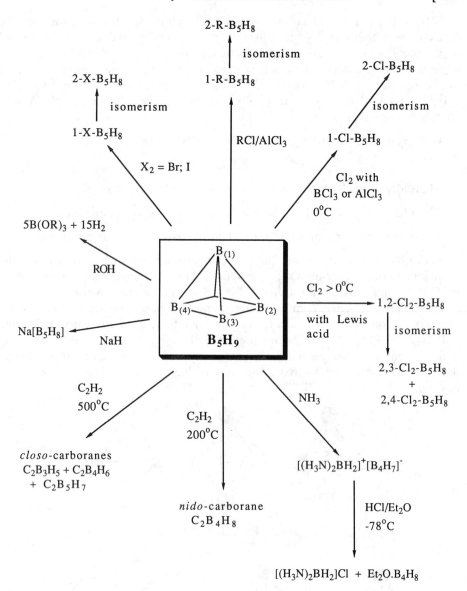

Fig. 7.1.5 — Some typical reactions of B_5H_9.

Some representative reactions of pentaborane(9) are summarized in Fig. 7.1.5. Hydrolysis by water to form boric acid is very slow, but with alcohols, ROH, complete degradation of the pentaborane(9) cluster occurs. Note that the alcohol reacts with the B–H bond as follows:

$$B-H + R-O-H \rightarrow B-OR + H_2$$

e.g. $B_5H_9 + 15\,MeOH \rightarrow 5B(OMe)_3 + 12H_2$

thus forming an orthoborate rather than boric acid. With less than the above stoichiometric amount of alcohol, the alkoxyboranes, $H_2B(OR)$ and $HB(OR)_2$, will be produced. Distinguishing an orthoborate from an alkoxyborane is readily achieved by using ^{11}B NMR spectroscopy; for example, a singlet resonance is observed for $B(OMe)_3$, a doublet for $HB(OMe)_2$, and a triplet for $H_2B(OMe)$, with coupling arising from ^1H–^{11}B spin–spin interactions.

The Friedel–Crafts alkylation of B_5H_9 results initially in the formation of an apically substituted B_5-cage. The apical atom, B(1), in pentaborane(9) is more negatively charged than the four basal boron atoms, and hence is the site at which electrophilic attack occurs. On heating at 200°C for several hours, the apically alkylated cage (i.e. the kinetic product) rearranges to a basally substituted product (i.e. the thermodynamic product). Similarly, halogenation of B_5H_9 initially gives the apically substituted 1-X-B_5H_8. The rearrangement of 1-X-B_5H_8 to 2-X-B_5H_8 is catalysed by base, and two pathways, the diamond–square–diamond or the base-swing mechanisms, are feasible (Fig. 7.1.6). The crucial feature, proven by ^{10}B labelling studies, is that dissociation of the substituent, X (X = alkyl or halide), does not occur during rearrangement. Rather, the B_5-cage itself undergoes a structural change. Contrasting with these electrophilic substitutions is the photolysis of B_5H_9 with $(CF_3)_2CO$ which leads to 2–{HO–C(CF_3)_2}–B_5H_8 and 1–{HO–C(CF_3)_2}–B_5H_8 in a 3:1 ratio.

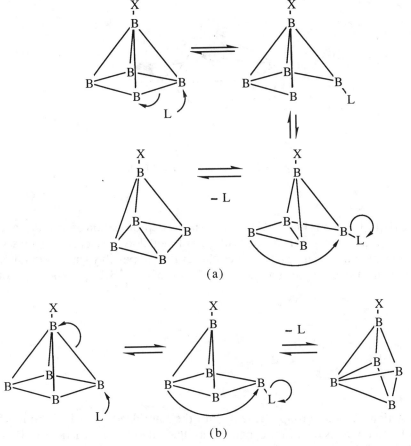

Fig. 7.1.6 — Rearrangement of 1–X–B_5H_8 to 2–X–B_5H_8 via (a) diamond–square–diamond, or (b) base-swing mechanism.

The reaction of pentaborane(9) with acetylenes leads to *closo-* or *nido-*carboranes depending upon the reaction conditions (Fig. 7.1.5). In contrast to the *closo*-carboranes described above, the isomer of the *nido*-carborane $C_2B_4H_8$ formed by pyrolysing B_5H_9 with C_2H_2 has the two carbon atoms in adjacent sites. The structure of *nido*-$C_2B_4H_8$ is related to that of B_6H_{10}. The two *endo*-hydrogen atoms bridge B–B edges of the open face of the cage (Fig. 7.1.7). This is a general phenomenon; as yet, there is no precedence for a C–H–B or C–H–C bridge. Pentaborane(9) is also a valuable precursor to higher boron hydrides, for example the $[B_9H_{14}]^-$ anion (see Fig. 4.9.2), $B_{10}H_{14}$, and $B_{18}H_{22}$ (see Fig. 4.13.7). The fusion of two pentaborane cages to give *conjuncto*-$\{B_5H_8\}_2$ (see Figs 4.13.1 and 4.13.2) is induced by electrical discharge. This *conjuncto*-cluster is stable *in vacuo* to 170°C, but, in the presence of iodine, rearranges at 150°C to *nido*-$B_{10}H_{14}$. The B–B σ-bond which joins the two B_5-sub-units in $\{B_5H_8\}_2$ is cleaved by reaction with HI:

$$\{B_5H_8\}_2 + HI \rightarrow B_5H_9 + B_5H_8I$$

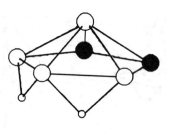

Fig. 7.1.7 — Structure of *nido*-$C_2B_4H_8$.

The $[B_5H_8]^-$ anion (Fig. 4.5.1) formed via the deprotonation of pentaborane(9), may be used as a precursor for the synthesis of metalloborane clusters (see section 7.4). Transition metals may also induce cluster fusion, possibly via a metalloborane intermediate; for example, the oxidative fusion of two $[B_5H_8]^-$ anions produces $B_{10}H_{14}$:

$$2[B_5H_8]^- \xrightarrow[\text{thf; } -78°C]{RuCl_3} B_{10}H_{14}$$

Insertion by main group elements into the unbridged basal B–B bond of the $[B_5H_8]^-$ anion leads to cluster expansion as illustrated by the reactions of $[B_5H_8]^-$ with Ph_2CHPCl_2 or Me_2NCH_2I:

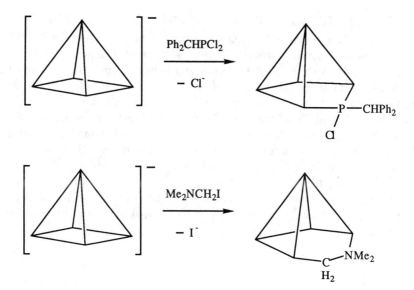

The formal reduction of pentaborane(9) leads to pentaborane(11), and in accord with Wade's rules, the addition of 2 hydrogen atoms (i.e. a source of 2 extra cluster bonding electrons) results in cage opening. The *arachno*-B_5H_{11} cluster is less stable than *nido*-B_5H_9, and fragmentation into B- and B_4-units is a common feature of its reactions. Hydrolysis is incomplete, even over a period of several hours at $\approx 90°C$:

$$B_5H_{11} + 3H_2O \rightarrow B_4H_{10} + B(OH)_3 + H_2$$

although, on acidification, degradation to boric acid does occur. The reactivity pattern of B_5H_{11} is typical of the small, open-cage boranes: reaction with Lewis bases cleaves the B_5-cage, alkylation occurs via a Friedel–Crafts pathway, and reaction with acetylenes leads to carborane clusters.

On the whole, the general chemistry of the hexaboranes, *nido*-B_6H_{10} and *arachno*-B_6H_{12}, reflects that of *nido*-B_5H_9 and *arachno*-B_5H_{11} respectively. However, in contrast to pentaborane(9), the presence of the unique unbridged basal B–B bond in hexaborane(10) provides this neutral cluster with an active site of basicity:

$$B_6H_{10} + HCl \rightarrow [B_6H_{11}]Cl$$

Hexaborane(10) also functions as an acid, and deprotonation by LiMe, NaH or KH yields the $[B_6H_9]^-$ anion, which, like $[B_5H_8]^-$, is a precursor to metalloborane clusters. Metal-induced oxidative fusion leads to $B_{12}H_{16}$:

$$2[B_6H_9]^- \xrightarrow[\text{Me}_2\text{O}; \ -78°C]{\text{FeCl}_3} B_{12}H_{16}$$

Hydrolysis of B_6H_{12} produces B_4H_{10} and $B(OH)_3$, whilst reaction with ether, R_2O, gives $R_2O.BH_3$ and B_5H_9.

7.2   THE CHEMISTRY OF $B_{10}H_{14}$: A LARGE, OPEN-CAGE BORANE

Decaborane(14) is a crystalline solid at room temperature (m.p. = 99.5°C; b.p. = 231°C); it does not spontaneously hydrolyse in neutral aqueous solution and is relatively stable with respect to aerial oxidation. Representative reactions of *nido*-$B_{10}H_{14}$ are illustrated in Fig. 7.2.1. The open face of the cluster accommodates four *endo*-hydrogen atoms (see Fig. 4.10.1). The conjugate base, $[B_{10}H_{13}]^-$, forms readily via the loss of one of the B–H–B protons, and, in the presence of a strong base such as NaH, deprotonation to the $[B_{10}H_{12}]^{2-}$ dianion may be effected. The *nido*-$B_{10}H_{14}$ cluster can accommodate two more electrons without gross structural change (see section 6.5); reduction is achieved by using sodium in liquid ammonia.

The charge distribution in decaborane(14) is such that boron atoms (1), (2), (3) and (4) (Fig. 7.2.1) possess the greatest share of the available electron density. These are the atoms of highest connectivity. Atoms B(6) and B(9) exhibit the lowest

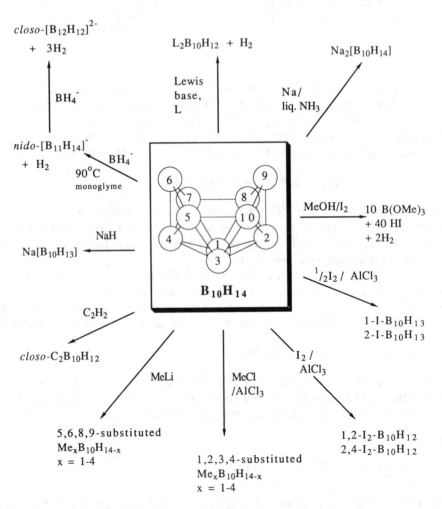

Fig. 7.2.1 — Some representative reactions of $B_{10}H_{14}$.

connectivity and possess correspondingly low electron density. This distinction is apparent in the products of electrophilic and of nucleophilic substitution. Friedel–Crafts alkylations give the 1-, 2-, 3-, and 4-substituted derivatives, while reaction with Me^- results in the formation of 6- and 9-substituted alkyldecaboranes; sites B(5) and B(8) are also susceptible towards nucleophilic attack. The introduction of a halogen via electrophilic substitution gives the 1-, 2- and 4-substituted $B_{10}H_{14-y}X_y$ ($y = 1$ or 2). If the reader is worried about any inconsistencies in these substitution patterns (e.g. why atom B(7) does not appear to behave in an analogous manner to the symmetry-related atom B(5)) then consider the effect on the charge distribution within the cage once the first substituent has been introduced.

Reactions of $B_{10}H_{14}$ with Lewis bases contrast with those described for the small boranes in which we saw that cluster fragmentation was a characteristic pathway. In the reaction

$$B_{10}H_{14} + 2Me_2S \rightarrow 6,9\text{-}B_{10}H_{12}(Me_2S)_2 + H_2$$

the $B_{10}H_{14}$ cage is essentially reduced; loss of two hydrogen atoms and a gain of two 2-electron donor ligands correspond to a net gain of 2 cluster electrons. Weak Lewis bases such as Me_2S are readily displaced by, for example, MeCN, Et_3As, Et_3N, or Ph_3P. Again, the B_{10}-cage remains intact during attack by Lewis base.

Reactions which expand the $B_{10}H_{14}$ cluster occur with boron hydrides, borane anions, acetylenes, and metal fragments. At ambient temperatures, $[BH_4]^-$ reduces $B_{10}H_{14}$ to $nido\text{-}[B_{10}H_{14}]^{2-}$, but at 90°C, the $nido\text{-}[B_{11}H_{14}]^-$ anion (Fig. 4.11.1) is produced. Further reaction with $[BH_4]^-$ leads to $closo\text{-}[B_{12}H_{12}]^{2-}$, via the insertion of a BH fragment with concurrent loss of molecular hydrogen (Fig. 7.2.1). The passage of a 1700 V a.c. discharge through decaborane(14) vapour and hydrogen gas yields the fused cluster, $B_{20}H_{16}$ (see Fig. 4.13.9). Acetylene inserts into the open face of the $B_{10}H_{14}$ cage, to generate the $closo$-carborane, $1,2\text{-}C_2B_{10}H_{12}$, as illustrated in Fig. 7.2.2. The cluster rearranges at 470°C to the 1,7-isomer, and at 700°C, to $1,12\text{-}C_2B_{10}H_{12}$, as illustrated in Fig. 7.2.3. The mechanism for each isomerism is either one of concerted bond breaking and forming (i.e. a diamond–square–diamond pathway related to that shown in Fig. 7.1.6) or a concerted twisting of the two pentagonal bipyramidal sub-units of the icosahedron.

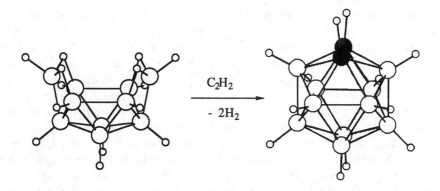

Fig. 7.2.2 — Conversion of $nido\text{-}B_{10}H_{14}$ to $closo\text{-}C_2B_{10}H_{12}$ by the insertion of acetylene.

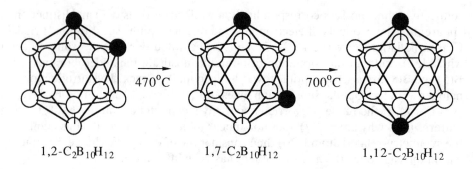

$$1,2\text{-}C_2B_{10}H_{12} \qquad 1,7\text{-}C_2B_{10}H_{12} \qquad 1,12\text{-}C_2B_{10}H_{12}$$

Fig. 7.2.3 — Isomerization of $1,2\text{-}C_2B_{10}H_{12}$ through to $1,12\text{-}C_2B_{10}H_{12}$.

7.3   THE CHEMISTRY OF THE *closo*-$[B_nH_n]^{2-}$ ANIONS

The *closo*-$[B_nH_n]^{2-}$ dianions exist for $6 \leqslant n \leqslant 12$. The preparative routes to these hydroborate anions tend to be non-selective: pyrolysis of $[B_3H_8]^-$ *in vacuo* produces $[B_9H_9]^{2-}$, $[B_{10}H_{10}]^{2-}$ and $[B_{12}H_{12}]^{2-}$; $[Et_4N][BH_4]$ reacts in refluxing decane–dodecane (175–190°C for 12 h) to give a mixture of $[B_9H_9]^{2-}$, $[B_{10}H_{10}]^{2-}$, $[B_{12}H_{12}]^{2-}$ and $[B_{11}H_{14}]^-$; the reaction of diborane with borohydride in a 1:1 stoichiometry produces the $[B_6H_6]^{2-}$ anion, although variation in the molar ratio of the reagents leads to higher hydroborate dianions. The $[B_7H_7]^{2-}$ and $[B_8H_8]^{2-}$ dianions may be obtained by oxidizing $[B_9H_9]^{2-}$.

The $[B_6H_6]^{2-}$, $[B_{10}H_{10}]^{2-}$ and $[B_{12}H_{12}]^{2-}$ clusters are thermally the most stable of the hydroborate dianions. However, this property is cation dependent; for example, $Cs_2[B_6H_6]$ is stable to $\approx 600°C$, whereas $Ag_2[B_6H_6]$ detonates on heating. The $[B_{10}H_{10}]^{2-}$ and $[B_{12}H_{12}]^{2-}$ dianions are also remarkably resistant towards hydrolysis, being kinetically stable in both acidic or alkaline solutions. Consequently, these two hydroborate dianions have been the most extensively studied. Large cations are required to stabilize the $[B_{10}H_{10}]^{2-}$ and $[B_{12}H_{12}]^{2-}$ dianions, and with Cu^+, Ag^+, and Hg^{2+}, there is evidence for extensive ion pairing. In the solid state, $Cu_2[B_{10}H_{10}]$ exhibits weak bonding interactions between each Cu^+ ion and an adjacent pair of $B_{(apical)}$ and $B_{(equatorial)}$ atoms.

Some representative reactions of $[B_{10}H_{10}]^{2-}$ and $[B_{12}H_{12}]^{2-}$ are shown in Figs 7.3.1 and 7.3.2 respectively. The chemistry tends to be dominated by electrophilic substitution, although some reactions with nucleophiles do occur. The icosahedral $[B_{12}H_{12}]^{2-}$ cluster possesses an even distribution of electron density, and hence, there is no preference for the initial site of electrophilic substitution. It is quite realistic to describe the $[B_{12}H_{12}]^{2-}$ anion as an inorganic aromatic system, since the delocalization of electron density over the borane cluster results in a reactivity pattern which is reminiscent of that of benzene. In the $[B_{10}H_{10}]^{2-}$ cage, there are two sites available for electrophilic substitution, and, as in B_5H_9, it is the apical boron atom which carries the greatest negative charge. The relative rates of deuteration for $B_{(apical)} : B_{(equatorial)}$ in $[B_{10}H_{10}]^{2-}$ is $\approx 4.9 : 1$. Preferential electrophilic substitution at the apical sites in $[B_{10}H_{10}]^{2-}$ is observed in the formation of the bis-diazo-derivative, $1,10\text{-}(N_2)_2B_{10}H_8$. However, the pattern is broken in, for example,

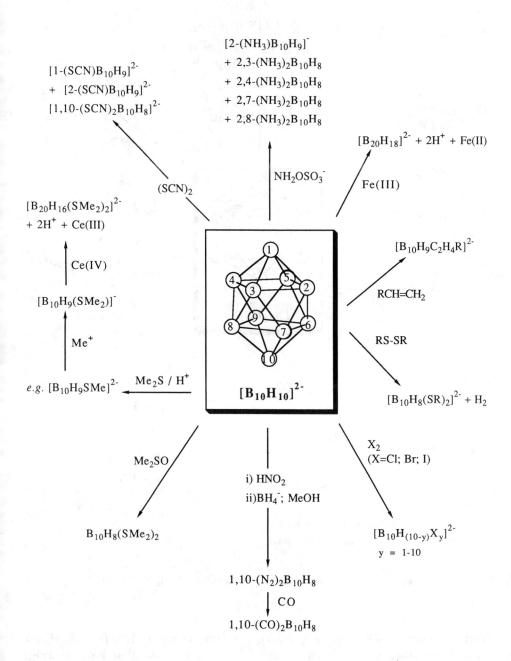

[2-(NH$_3$)B$_{10}$H$_9$]$^-$
+ 2,3-(NH$_3$)$_2$B$_{10}$H$_8$
+ 2,4-(NH$_3$)$_2$B$_{10}$H$_8$
+ 2,7-(NH$_3$)$_2$B$_{10}$H$_8$
+ 2,8-(NH$_3$)$_2$B$_{10}$H$_8$

[1-(SCN)B$_{10}$H$_9$]$^{2-}$
+ [2-(SCN)B$_{10}$H$_9$]$^{2-}$
[1,10-(SCN)$_2$B$_{10}$H$_8$]$^{2-}$

[B$_{20}$H$_{18}$]$^{2-}$ + 2H$^+$ + Fe(II)

NH$_2$OSO$_3^-$

Fe(III)

(SCN)$_2$

[B$_{20}$H$_{16}$(SMe$_2$)$_2$]$^{2-}$
+ 2H$^+$ + Ce(III)

Ce(IV)

[B$_{10}$H$_9$(SMe$_2$)]$^-$

[B$_{10}$H$_9$C$_2$H$_4$R]$^{2-}$

RCH=CH$_2$

Me$^+$

RS-SR

e.g. [B$_{10}$H$_9$SMe]$^{2-}$ Me$_2$S / H$^+$

[B$_{10}$H$_{10}$]$^{2-}$

[B$_{10}$H$_8$(SR)$_2$]$^{2-}$ + H$_2$

Me$_2$SO

i) HNO$_2$
ii)BH$_4^-$; MeOH

X$_2$
(X=Cl; Br; I)

B$_{10}$H$_8$(SMe$_2$)$_2$

[B$_{10}$H$_{(10-y)}$X$_y$]$^{2-}$
y = 1-10

1,10-(N$_2$)$_2$B$_{10}$H$_8$

CO

1,10-(CO)$_2$B$_{10}$H$_8$

Fig. 7.3.1 — Some representative reactions of the [B$_{10}$H$_{10}$]$^{2-}$ dianion.

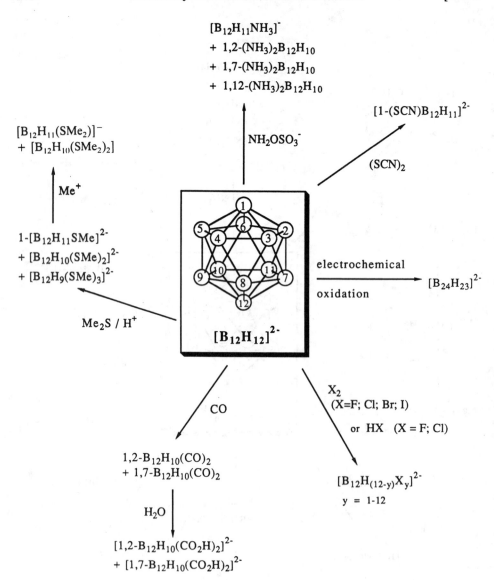

Fig. 7.3.2 — Some representative reactions of the $[B_{12}H_{12}]^{2-}$ dianion.

amination of the cluster (see Fig. 7.3.1). The diazo-groups in $1,10-(N_2)_2B_{10}H_8$ are readily displaced by Lewis bases such as CO. Sulphur–sulphur bond cleavage accompanies the reactions of $[B_{10}H_{10}]^{2-}$ or $[B_{12}H_{12}]^{2-}$ with thioethers or thiocyanogen, $(SCN)_2$.

Direct halogenation of the $[B_{10}H_{10}]^{2-}$ and $[B_{12}H_{12}]^{2-}$ anions occurs with Cl_2, Br_2 or I_2 in aqueous or alcoholic solution, and the rate of reaction follows the order $Cl > Br > I$. The derivatives have the general formula $[B_{10}H_{(10-y)}X_y]^{2-}$ or

$[B_{12}H_{(12-y)}X_y]^{2-}$; complete halogenation may be achieved, but the rate of substitution decreases as the number of substituents increases.

Oxidative cluster fusion may be achieved either chemically or electrochemically. Coupling two icosahedral cages (probably via a B–H–B bridge) yields the $[B_{24}H_{23}]^{3-}$ ion, but only with an electrochemical route; chemical oxidation degrades the cluster with concurrent boron–oxygen bond formation. Iron(III) ions oxidize $[B_{10}H_{10}]^{2-}$ to $[B_{20}H_{18}]^{2-}$; the bicapped antiprismatic sub-units remain intact, fused together via a four-centre interaction involving an apical and an equatorial boron atom on each cage.

The hydrolytic stability of the $[B_nH_n]^{2-}$ anions varies with n as follows: $12 > 10 \gg 11 > 9 \approx 8 \approx 6 > 7$. Thus, the significantly lower hydrolytic stability, coupled with their decreased stability with respect to oxidation, has mitigated against the development of the chemistry of dianions other than $[B_{10}H_{10}]^{2-}$ and $[B_{12}H_{12}]^{2-}$. Under controlled moisture- and oxygen-free conditions, substitutions do occur. Halogenation is the most commonly exploited reaction; e.g. chlorination of $[B_8H_8]^{2-}$ yields $[B_8H_{8-y}Cl_y]^{2-}$ ($y = 1$–8); $[B_9H_3Br_6]^{2-}$, $[B_9Br_9]^{2-}$, $[B_{11}H_2Br_9]^{2-}$ and $[B_{11}H_7Br_4]^{2-}$ are characterized products from the bromination of $[B_9H_9]^{2-}$ and $[B_{11}H_{11}]^{2-}$. Other substitutions include

$$[B_9H_9]^{2-} + NH_2OSO_3H \rightarrow [B_9H_8(NH_3)]^- + HSO_4^-$$

$$[B_9H_8(NH_3)]^- + 3MeI \rightarrow [B_9H_8(NMe_3)]^- + 3HI$$

$$[B_{11}H_{11}]^{2-} + (MeCO)_2O + Me_2SO \rightarrow [B_{11}H_{10}(SMe_2)]^- + MeCO_2^- + MeCO_2H$$

Substitution in $[B_9H_9]^{2-}$ occurs preferentially at an apical boron atom as expected.

7.4   THE INTERACTION OF BORANES WITH METAL ATOMS

Although a high proportion of metalloborane compounds are produced by serendipitous routes, the rational syntheses which are documented provide us with a useful indication of the reactivity patterns of their borane precursors. One simple strategy is to replace a proton by a transition metal-based electrophile:

$$B_xH_y \xrightarrow{-H^+} [B_xH_{y-1}]^- \xrightarrow{[ML_z]^+} L_zMB_xH_{y-1}$$

This general scheme is used widely in the synthesis of metalloboranes, and the source of $[ML_z]^+$ is often a transition metal halide. Each of the reactions following is conveniently regarded as a nucleophilic substitution at the metal centre:

$$(Ph_3P)_3CuCl + [B_5H_8]^- \rightarrow (Ph_3P)_2CuB_5H_8 + Cl^- + PPh_3$$

$$(Ph_3P)_3CuCl + [B_6H_9]^- \rightarrow (Ph_3P)_2CuB_6H_9 + Cl^- + PPh_3$$

$$(cyclo\text{-}C_6H_{11})_3PAuCl + [B_{10}H_{13}]^- \rightarrow (cyclo\text{-}C_6H_{11})_3PAuB_{10}H_{13} + Cl^-$$

Reactions of the $[B_3H_8]^-$ anion may be considered as a special case, since the conjugate acid, 'B_3H_9', has not been isolated:

$$(\eta^5\text{-Cp})Fe(CO)_2I + [B_3H_8]^- \rightarrow (\eta^5\text{-Cp})Fe(CO)_2B_3H_8 + I^-$$

$$BeCl_2 + 2[B_3H_8]^- \rightarrow Be(B_3H_8)_2 + 2Cl^-$$

$$(CO)_5MnBr + [B_3H_8]^- \rightarrow (CO)_4MnB_3H_8 + Br^- + CO$$

$$(dppe)PdCl_2 + [B_3H_8]^- \rightarrow (dppe)Pd(\eta^3\text{-}B_3H_7) + Cl^- + HCl$$

dppe = $Ph_2PCH_2CH_2PPh_2$

Note that in the last reaction, rearrangement of the triborane ligand occurs, presumably as the second equivalent of chloride ion is expelled. This contrasts with the simple halide ion displacement in the reaction with $BeCl_2$.

Substitution of one or more boron atoms by metal fragments in a cluster is a commonly exploited synthetic route, and is readily exemplified by using some reactions of pentaborane(9). In many cases, the reaction products are difficult to rationalize, and yields of a given product are often low. Reaction conditions are usually critical:

$$Co(vap) + C_5H_6 + B_5H_9 \xrightarrow{-196°C} \{(\eta^5\text{-Cp})Co\}_3B_5H_5 + \{(\eta^5\text{-Cp})Co\}_3B_5H_4\{\sigma\text{-}cyclo\text{-}C_5H_7\}$$

$$Fe(CO)_5 + LiAlH_4 + B_5H_9 \xrightarrow{OEt_2} Fe_2(CO)_6B_2H_6 \ (\leqslant 10\%)$$

$$Fe(CO)_5 + B_5H_9 \xrightarrow[220°C\text{-}20°C]{hot\text{-}cold\ reactor} 1\text{-}Fe(CO)_3B_4H_8 \ (10\text{-}20\%)$$

The introduction of a metal fragment may result in cage expansion rather than a substitution reaction:

$$Mn_2(CO)_{10} + B_5H_9 + H_2 \xrightarrow{140°C} 2\text{-}Mn(CO)_3B_5H_{10} \ (\approx 9\%)$$

Cluster expansion is also observed in some reactions of borane anions with transition metal compounds, thus contrasting with the idea of introducing a metal fragment as a replacement for an *endo*-hydrogen atom. The reaction pathway depends upon the bonding abilities of the metal fragment. If, as in $[AuPR_3]^+$, the frontier orbital set is limited to an acceptor orbital of σ-symmetry (i.e. analogous to a proton) then a simple bridging mode will probably be adopted. If, as in $\{Fe(CO)_3\}$ or $\{Ir(CO)_3\}$, the metal fragment has frontier orbitals which are comparable to those of a cluster BH unit (i.e. orbitals of both σ- and π-symmetry) then cluster expansion is likely to occur (see Chapter 6). In the reaction

$$[B_4H_9]^- + trans\text{-}IrCl(CO)(PMe_2Ph)_2 \rightarrow 1\text{-}Ir(CO)(PMe_2Ph)_2B_4H_9$$

the structure of the product is related to that of B_5H_{11} with the iridium fragment residing in the apical position. In the reaction

$$[B_4H_9]^- + RhCl(PPh_3)_3 \rightarrow 2\text{-}Rh(H)(PPh_3)_2B_4H_8$$

the product possesses 2 fewer cluster bonding electrons than $1\text{-}Ir(CO)(PMe_2Ph)_2B_4H_9$, and therefore exhibits a structure which is related to B_5H_9 rather than B_5H_{11}. In $1\text{-}Ir(CO)(PMe_2Ph)_2B_4H_9$, the tetraborane may be described as an η^4-ligand, whilst in $2\text{-}Rh(H)(PPh_3)_2B_4H_8$, it functions as an η^3-ligand towards the metal fragment. Some related reactions are shown below:

$$[B_5H_8]^- + trans\text{-}IrCl(CO)(PPh_3)_2 \xrightarrow{-25°C} 2\text{-}Ir(CO)(PPh_3)_2B_5H_8$$

$$[B_9H_{14}]^- + trans\text{-}IrCl(CO)(PPh_3)_2 \rightarrow [6\text{-}Ir(H)(PPh_3)_2B_9H_{13}]$$

$$[B_9H_{14}]^- + Re(CO)_5Br \rightarrow [6\text{-}Re(CO)_3B_9H_{13}]^-$$

$$2[B_{10}H_{13}]^- + PdCl_2(cod) \xrightarrow{MeCN/thf} [7,7'\text{-}Pd(B_{10}H_{12})_2]^{2-}$$
cod = *cyclo*-octadiene

$$[B_{11}H_{13}]^{2-} + (\eta^5\text{-}Cp)_2Ni \xrightarrow[Na/Hg]{MeCN} [1\text{-}(\eta^5\text{-}Cp)NiB_{11}H_{11}]^-$$

Reactions of the *closo*-hydroborate dianions with transition metal fragments lead to a variety of novel compounds. Possible pathways are:

(1) insertion of the metal into the borane cage leading to cage expansion;
(2) substitution of a metal fragment for a BH unit;
(3) cluster degradation;
(4) association of the metal fragment with the *exo*-hydrogen atoms of the *closo*-cage.

In the reaction of $[B_{10}H_{10}]^{2-}$ with $RhCl_3(PMe_2Ph)_3$, metal-for-boron atom substitution competes with metal insertion, yielding several *closo*-rhodaboranes of differing cage size. In an ethanol/chloroform solution, $[B_{10}H_{10}]^{2-}$ reacts with $RuCl_2(PPh_3)_3$ to give $Ru(PPh_3)_2B_{10}H_8(OEt)_2$, in which two terminal H atoms have been replaced by OEt substituents. The eleven-atom cage of $Ru(PPh_3)_2B_{10}H_8(OEt)_2$ is structurally analogous to that of $[B_{11}H_{11}]^{2-}$; the metal atom occupies the unique, 6-coordinate site (see Fig. 4.1.1). $Ru(PPh_3)_2B_{10}H_8(OEt)_2$ reacts further with a second metal fragment:

$$Ru(PPh_3)_2B_{10}H_8(OEt)_2 + MCl_2(PPh_3)_3 \rightarrow$$

$$\{(PPh_3)_2MCl\}ClRu(PPh_3)_2B_{10}H_8(OEt)_2$$
$$M = Ru; Os$$

In the product cluster, the newly added $\{(PPh_3)_2MCl\}$-fragment is held in an *exo*-site relative to the B_{10}-cage via an Ru–Cl–M and two B–H–M bridge interactions. A related mode of attachment is observed for the $\{(PPh_3)_2RhCl\}$-fragment in $\{(PPh_3)_2RhCl\}B_{12}H_{11}(NEt_3)$:

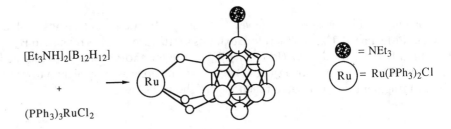

$[Et_3NH]_2[B_{12}H_{12}]$

+

$(PPh_3)_3RuCl_2$

⬤ = NEt₃

Ru = Ru(PPh₃)₂Cl

7.5   REACTIVITY OF METALLOBORANES

A discussion of the reactivity of metalloborane clusters is conveniently considered in two sections: (i) boron-rich metalloboranes, and (ii) metal-rich metalloboranes. How does the reactivity of a boron-rich metalloborane differ from that of the parent boron hydride? If we compare B_5H_9 with $1\text{-Fe(CO)}_3B_4H_8$, $1\text{-}(\eta^5\text{-Cp})CoB_4H_8$, or $2\text{-}(\eta^5\text{-Cp})CoB_4H_8$, then, clearly, the charge distribution over the cluster atoms will have changed, and with it the acidity of the *endo*-hydrogen atoms. In addition, in going from B_5H_9 to $2\text{-}(\eta^5\text{-Cp})CoB_4H_8$, two of the four *endo*-hydrogen atoms are transformed from B–H–B to Co–H–B bridging modes. Deprotonation of a metalloborane cluster tends to occur via loss of a B–H–M rather than a B–H–B proton. Anions so formed are useful reagents for cluster expansion reactions; hence the general reaction of a borane anion with a metal fragment to give a metalloborane cluster may be extended to incorporate several metal atoms. For example:

$$[2\text{-}(\eta^5\text{-Cp})CoB_4H_7]^- + CoCl_2 + Na(C_5H_5) \rightarrow 1,2\text{-}\{(\eta^5\text{-Cp})Co\}_2B_4H_6$$
$$+ 1,2,3\text{-}\{(\eta^5\text{-Cp})Co\}_3B_4H_4$$
$$+ 1,2,3\text{-}\{(\eta^5\text{-Cp})Co\}_3B_3H_5$$
$$+ 3,4,5,6\text{-}\{(\eta^5\text{-Cp})Co\}_4B_4H_4$$

A second metal atom may be inserted into a neutral metalloborane cluster. The preparation of $\{(PPh_3)_2MCl\}ClRu(PPh_3)_2B_{10}H_8(OEt)_2$ (M = Ru; Os) was described above. The elimination of hydrogen halide is often a driving force for such a reaction:

$$arachno\text{-}(Me_3P)_2(CO)HIrB_8H_{12} + cis\text{-}(PMe_3)_2PtCl_2 \rightarrow$$
$$\{(PMe_3)_2Pt\}\{Me_3P)_2(CO)HIr\}B_8H_{10}$$

Cluster rearrangements constitute a significant part of the chemistry of boron-rich metalloboranes. One example was illustrated in Fig. 5.7.3, in which cage opening (or closing) accompanies formal reduction (or oxidation) of the cluster. The observed reactions are

$$arachno\text{-}(CO)H(PMe_3)_2IrB_8H_{12} \rightarrow nido\text{-}(CO)(PMe_3)_2IrB_8H_{11} + H_2$$

$$nido\text{-}(CO)(PMe_3)_2IrB_8H_{10}Cl \rightarrow closo\text{-}H(PMe_3)_2IrB_8H_7Cl + H_2 + CO$$

The final product possesses 9 pairs of cluster bonding electrons and yet, unpredictably, exhibits a closed-cage structure with 9 vertices. For this reason, it is termed an *isocloso*-cluster.

The chemistry of metal-rich metalloboranes is complicated by the potential for competition between the reactivity of the transition metal and boron atoms. An instructive example is the reaction of $[HFe_3(CO)_9BH_3]^-$ with a Lewis base such as a phosphine. The boron atom is susceptible to attack by Lewis base, but then, transition metal carbonyl compounds readily undergo substitution reactions with phosphines. In addition, both multinuclear transition metal clusters and boranes are often cleaved as a result of attack by a nucleophile. The observed reaction of $[HFe_3(CO)_9BH_3]^-$ with a phosphine such as $PhMe_2P$ is a series of competitive pathways as illustrated in Fig. 7.5.1. At low concentrations of phosphine, substitution of metal-bound CO for PR_3 is preferred, whilst at high concentrations of phosphine, the Fe_3B-cluster is cleaved, either into Fe_2B + Fe units or into Fe_3 + B units.

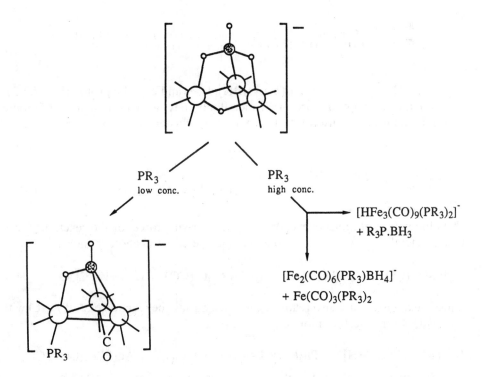

Fig. 7.5.1 — Competitive pathways in the reaction of $[HFe_3(CO)_9BH_3]^-$ with the phosphine, PR_3 (R = alkyl or aryl).

The deprotonation of a metal-rich metalloborane usually occurs via the loss of an M–H–B proton, but this is not a general rule. For example, whilst reaction with a base removes an Fe–H–B proton from $(\mu\text{-H})Fe_3(CO)_9BH_4$, the closely related

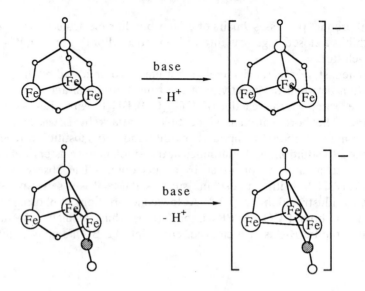

Fig. 7.5.2 — Comparison of the sites of deprotonation of $(\mu\text{-H})Fe_3(CO)_9BH_4$ and $(\mu\text{-H})Fe_3(CO)_9(\mu\text{-CO})BH_2$. Each Fe atom carries 3 terminal carbonyl ligands.

cluster, $(\mu\text{-H})Fe_3(CO)_9(\mu\text{-CO})_9(\mu\text{-CO})BH_2$, loses an Fe–H–Fe proton (Fig. 7.5.2). As with the boranes or boron-rich metalloboranes, a metal-rich metalloborane cluster anion may be a useful starting point for a cluster expansion reaction, e.g.

$$[(\mu\text{-H})Fe_3(CO)_9BH_3]^- + 2Fe_2(CO)_9 \xrightarrow[\text{toluene}]{\text{room temp}}$$
$$[(\mu\text{-H})Fe_4(CO)_{12}BH]^- + H_2 + 3Fe(CO)_5$$

The elimination of molecular hydrogen is one driving force for the reaction. For a neutral metalloborane, a similar reaction may be initiated photochemically:

$$(\mu\text{-H})Ru_3(CO)_9BH_4 + Fe(CO)_5 \xrightarrow{h\nu} (\mu\text{-H})Ru_3Fe(CO)_{12}BH_2 + H_2 + 2CO$$

Cluster expansions are also readily achieved by using the reaction of $AuPPh_3Cl$ with a metalloborane cluster anion, e.g.

$$[(\mu\text{-H})Fe_4(CO)_{12}BH]^- + Ph_3PAuCl \rightarrow (\mu-H)Fe_4(CO)_{12}Au(PPh_3)BH + Cl^-$$

$$[(\mu\text{-H})Fe_4(CO)_{12}BH]^- + 2Ph_3PAuCl \rightarrow Fe_4(CO)_{12}\{Au(PPh_3)\}_2BH + 2Cl^-$$
$$+ H^+$$

$$[(\mu\text{-H})Fe_4(CO)_{12}BH]^- + 3Ph_3PAuCl \rightarrow Fe_4(CO)_{12}\{Au(PPh_3)\}_3B + 3Cl^- + 2H^+$$

A related reaction is

$$[(\mu\text{-H})Fe_4(CO)_{12}BH]^- + [Rh(CO)_2Cl]_2 \rightarrow [Fe_4Rh_2(CO)_{16}B]^- + 2HCl$$

REFERENCES

The following are selected references which include discussions of the reactivity of boranes and metalloboranes:

[1] J. F. Liebman, A. Greenberg and R. E. Williams, eds *Advances in Boron and the Boranes*, VCH Publishers, Weinheim, 1988.

[2] K. Wade, *Electron Deficient Compounds*, Nelson, London, 1971.

[3] N. N. Greenwood, *Pure Appl. Chem.*, 1977, **49**, 791.

[4] R. N. Grimes, *Pure Appl. Chem.*, 1982, **54**, 43.

[5] R. N. Grimes, in *Comprehensive Organometallic Chemistry*, eds G. Wilkinson, F. G. A. Stone and E. W. Abel, Volume 1, p. 459, Pergamon, Oxford, 1982.

[6] K. B. Gilbert, S. K. Boocock and S. G. Shore, in *Comprehensive Organometallic Chemistry*, eds G. Wilkinson, F. G. A. Stone and E. W. Abel, Volume 6. p. 879, Pergamon, Oxford, 1982.

[7] R. N. Grimes, ed. *Metal Interactions with Boron Clusters*, Plenum Press, New York, 1982.

[8] J. D. Kennedy, *Prog. Inorg. Chem.*, 1984, **32**, 519; *ibid.*, 1986, **34**, 211.

[9] C. E. Housecroft, *Polyhedron*, 1987, **6**, 1935.

[10] T. P. Fehlner, *New J. Chem.*, 1988, **12**, 307.

Formula index

Subject index

BC1